TEACH YOURSELF

D0528675

GEOLOGY
David A. Rothery

To the students of Earth Sciences
at the Open University, who taught me
how much fun geology can be.

Long-renowned as the authoritative source for self-guided learning – with more than 30 million copies sold worldwide – the *Teach Yourself* series includes over 200 titles in the fields of languages, crafts, hobbies, sports, and other leisure activities.

A catalogue record for this title is available from the British Library.

Library of Congress Catalog Card Number: 96-72379.

First published in UK 1997 by Hodder Headline Plc, 338 Euston Road, London NW1 3BH

First published in US 1997 by NTC Publishing Group, 4255 West Touhy Avenue, Lincolnwood (Chicago), Illinois 60646 – 1975 U.S.A.

The 'Teach Yourself' name and logo are registered trade marks of Hodder & Stoughton Ltd in the UK.

Typeset by Transet Limited, Coventry, England.
Printed in England by Cox & Wyman Limited, Reading, Berkshire.

Impression number   10 9 8 7 6 5   4 3
Year   2000   1999 1998

CONTENTS

ACKNOWLEDGEMENTS

I am immensely grateful to Jayne Walker, who read the entire first draft of this book, for her prompt and sound advice. I am also indebted to my friends and colleagues who so willingly took the time to discuss many specific points with me while the book was in preparation, and to my family for putting up with my cries of 'go away' when they attempted to interrupt me.

The diagrams in this book have been specially prepared, and the majority are brand new. Most of the photographs are my own. With those exceptions, the illustrations are from the following sources:

Figure 3.2 NOAA National Geophysical Data Center; Figure 5.6 Stuart Hall; Figure 5.12 Peter Francis; Figures 13.3–13.6 are modified from various sources, but draw heavily on the work of C. R. Scotese and co-workers; Figure 14.2 *Magellan* image courtesy of G. H. Pettengill and US National Space Science Data Center; Figure 14.5 *Galileo* image courtesy of M. Belton and NASA Jet Propulsion Laboratory; Plate 13 Steve Drury; Plate 14 John Watson (copyright Open University).

1

INTRODUCTION: A MIGHTY MATTER OF LEGEND

'The green earth, say you? That is a mighty matter of legend, though you tread it under the light of day!' Although taken from a work of fiction*, about a fictional situation, those words may be taken as the theme for this book. The Earth *is* a mighty matter of legend. Its origin and the events that have taken place in its history do indeed form a grand and epic tale. Moreover, various parts of this story have been interpreted in many ways; some of these are contradictory, and others are now thought to be untrue.

I should also point out that some of what Earth scientists now believe fairly firmly will undoubtedly turn out to be wrong, or at best only partly correct. However, I think it is fair to claim that we understand very well (for the purpose of an introductory book such as this) most of the processes that shape the materials at and immediately beneath the Earth's surface. Probably, what we lack most is a fully developed understanding of the interdependence of all these processes. Even so we now have a fairly complete understanding of how the Earth as a whole functions, at least in broad outline.

People may become interested in geology, which is the scientific study of the Earth, for a variety of reasons. Many are intrigued by fossils, or pebbles that they pick up on the beach, and wonder what they mean. Others want to know more about earthquakes and volcanoes reported in the news, and whether such events are likely in their own neighbourhood. Some may be curious about where the oil comes from that powers their car, or what their house bricks are made of. And some

* Aragorn to Eomer, *The Two Towers*, J. R. R. Tolkien, 1954

are just overawed by the grandeur of the mountains or impressed by the pictures of other Earth-like bodies sent back by spaceprobes, and decide they want to find out more.

Whatever your own reason for opening this book, I hope you will discover what you want. I have tried to address all these concerns, and others. Most chapters aim to help you understand a particular type of geological process. You don't have to read these chapters in the sequence in which they are printed, but it will be easier if you do. This is because I have attempted to build up the use of terminology (and geology is riddled with strange sounding words!) in a coherent fashion. Notable exceptions to the process-orientated approach are the next chapter, which is a descriptive account of the composition of the Earth, and the final chapter, which deals with fieldwork and the simple equipment you will find useful if you decide to go and look at rocks and fossils for yourself. There is a Glossary at the end, where the most important terms (introduced in **bold** in the text) are defined.

It is helpful at this stage to define terms that describe each of the three main kinds of rocks. **Igneous** rocks are those which were once molten; they usually contain crystals that grew within this molten material as it cooled. **Metamorphic** rocks form when heat or pressure (or both) causes a pre-existing rock to recrystallise, but without melting. **Sedimentary** rocks are those rocks that form at the surface, by deposition of detrital grains or by precipitation from solution in water.

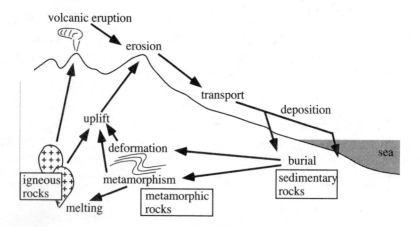

Figure 1.1 Cross-section through the Earth's surface to show the rock cycle.

A simplified view of origins of these rock types is shown in Figure 1.1, which illustrates what is known as the 'rock cycle'. This is the erosion, transport, deposition, burial, heating, deformation, melting, cooling, and exhumation of rocks or rock fragments. Rocks on high ground are worn to fragments by frost, rain and wind, and transported downhill to a place where these fragments get deposited and buried. This is the erosional and depositional part of the rock cycle. If these processes were the sole agent of change, the ultimate fate of the Earth's surface would be for hills and mountains to be worn away, until eventually everywhere was flat and buried by mud. However, the Earth is a dynamic planet and there are forces of deformation and uplift at work that continually bring igneous and metamorphic rocks, and hardened sediment, to the surface. This provides an unending supply of material to be worked on by the forces of erosion, and so the rock cycle continues.

In nature the rock cycle is not as simple as it may look in Figure 1.1. If we were to attempt to trace the history of a particular volume of rock over hundreds of millions of years, we might find it tracing a loop to and fro through just part of the rock cycle, perhaps being alternately metamorphosed and melted at depth and never being brought to the surface. On the other hand a grain of sediment may be deposited in a whole succession of sedimentary environments, being picked up by erosion and transported to each new environment without ever becoming deeply buried.

There is no obvious point at which to begin describing the rock cycle. In this book I begin with the igneous parts of the rock cycle, near the surface (Chapter 5) and at depth (Chapter 6). Next I deal with metamorphism (Chapter 7), followed by the erosional and depositional parts of the cycle (Chapters 8 and 9), and then the remaining aspect of the rock cycle, which is how rocks get deformed (Chapter 10). Subsequent chapters consider the physical resources that are available to us as a result of the rock cycle, the history of life on Earth and of the Earth itself, and evidence that the processes we have described occur on other planets too. The final chapter provides an introduction to geological fieldwork.

First, though, I describe the Earth's internal properties (Chapter 2), earthquakes (Chapter 3), which are one of the most obvious manifestations of the unquiet Earth, and plate tectonics (Chapter 4), which is a description of how wandering continents and widening or vanishing oceans are associated with the creation and destruction of rock on a huge scale.

2

THE SOLID EARTH

Before exploring the geological processes that affect the Earth, we had better establish just what we are dealing with. So, what is the Earth? As Table 2.1 shows, it is the largest of several rocky planets in our solar system. It is nearly spherical, measuring 12 714 km from pole to pole and 12 756 km across the Equator, and orbits the Sun at an average distance of 150 million km.

Name	Distance from Sun in millions of km	Equatorial radius in km	Mass, relative to Earth (6.0 x 10^{24} kg)	Density in tonnes per cubic metre (10^3 kg m^{-3})
Mercury	58	2439	0.055	5.43
Venus	108	6052	0.815	5.25
Earth	150	6378	1	5.52
Moon	150	1738	0.0123	3.34
Mars	228	3394	0.107	3.90

Table 2.1 The Earth and similar planets. Although the Moon orbits the Earth instead of the Sun it is included here because geologically it belongs to the same class of bodies.

The atmosphere and its development

Compared to the size of the Earth, the atmosphere is no more than a thin layer. It is so sparse as to be virtually unbreathable at the tops of the highest mountains (8–9 km above sea level), the highest common clouds occur at about 12 km, and it is insubstantial enough to offer little or no resistance to artificial satellites orbiting at as low as 200 km altitude. However, so far as living things are concerned, the importance of the atmosphere is out of all proportion to its size. This is not just because it contains the oxygen necessary for breathing, but also because the atmosphere moderates the temperature of the globe and shields the surface from those wavelengths of light that are harmful to life. Most familiar living things would have a hard time surviving on the Earth's surface if the atmosphere were to become radically changed.

This is not to say that the present atmosphere has been around since the Earth was formed. As we will see towards the end of this chapter, the Earth's origin is reliably dated at about 4550 million years ago. At that time the Earth's primordial atmosphere, most of which is thought to have escaped from the interior by means of volcanoes, was probably mostly water vapour, carbon dioxide, sulphur dioxide, and nitrogen. The appearance and subsequent evolution of exceedingly primitive living organisms (bacteria and simple single-celled plants) evidently changed the atmosphere, for example, liberating oxygen by the breakdown of carbon dioxide and sulphur dioxide. This, in turn, made it possible for higher organisms to arise.

The atmosphere seems to have had only about 1% of its present content of oxygen about 2000 million years ago, when the earliest known plant cells with nuclei appeared. With the arrival of the first land plants, about 500 million years ago, oxygen reached about one-third of its present concentration. It had risen to almost its present level by about 370 million years ago, when animals first emerged onto land. Today's atmosphere is thus as much a product of life as a requirement to sustain life as we know it.

The Earth's interior

This book is concerned mostly with things that go on at, or at shallow depth below, the Earth's surface. We will come back to the atmosphere from time to time, but first we need to investigate the deep interior. People living in regions where the soil is thin are well aware that the outside of the solid Earth is made of rocks. The rest of us are reminded of this when we see a quarry or a cliff. But is the Earth made of rocky material like this all the way to the centre?

Well, there is a simple way to find out, which is to consider the Earth's density. The kind of rock we usually find at the surface has a density of about 2.7 tonnes per cubic metre. If the Earth were made of rock, it should have an overall density not too different from this (perhaps a little greater, to allow for compression, and hence greater density, at depth).

We can work out the Earth's actual density as follows. The Earth's mass is known to a high degree of precision, because it can be determined from how long it takes the Moon or an artificial satellite to orbit the Earth. It is shown in Table 2.1 as 6.0×10^{24} kg. If you are not familiar with scientific notation, this translates as six million million million million kilogrammes, or six thousand million million million tonnes. To find the Earth's density, we just have to divide its mass by its volume. The result is 5.52 tonnes per cubic metre, which is about twice the density of surface rock. Clearly, the Earth is not just rock. If the outer part is rock, there must be something *denser* than the average value hidden away at depth. In fact, the centre of the Earth is believed to be composed mostly of iron, with a density of almost 13 tonnes per cubic metre.

The picture geologists have formed of the Earth's interior is not based merely on density arguments. All that density tells us is that there must be something much denser than rock somewhere within the Earth. It does not tell us how this denser material is arranged, let alone what it is made of. For example, Figure 2.1 shows two entirely different types of internal structure for the Earth, both of which are consistent with the Earth's density.

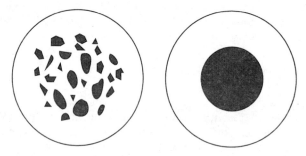

Figure 2.1 Two possible arrangements of dense material within the Earth. Either of these would give the observed density, but only the model on the right fits the observations discussed later in the text.

Seismic waves

The most useful information for determing the internal structure of the Earth comes from the study of how vibrations travel through it. As you can imagine, it needs pretty powerful vibrations to be detected through the full thickness of the globe. Fortunately, the vibrations from earthquakes, which are called **seismic waves**, are sufficiently powerful for this purpose (though this is a mixed blessing for people living in the most earthquake-prone areas, as we will see in the next chapter).

The reason why seismic waves are so useful for probing the Earth's interior is that the speed at which vibrations travel through a substance depends on the physical properties of that substance. As density increases, so the speed of seismic waves tends to decrease. On the other hand, vibrations travel faster through a more-rigid substance than through a less-rigid substance.

Let's see how this can help us determine the structure of the Earth, by supposing for now that the outer part of the Earth has a uniform (rocky) composition. How would we expect the speed of seismic waves to vary with depth? Rigidity should increase with depth, as a result of the ever-increasing pressure exerted by the overlying material. The effect of this should be for the speed of transmission of seismic waves to increase as depth increases. The contrary tendency for speed to decrease in response to increasing density with depth is much smaller, because rock is not compressed very much however high the pressure gets. We therefore expect the speed of seismic waves to increase with depth within the rocky part of the Earth, and this is indeed what we find.

We can tell this when an earthquake happens at a known location by timing the arrivals of the first vibrations received by seismic recording stations at different distances away. Figure 2.2 shows how this works. Note that the vibrations follow curved paths as a consequence of the speed of the transmission increasing with depth.

Studies of this sort, based on hundreds of thousands of earthquakes detected at thousands of seismic stations, has enabled the speed of seismic waves at different depths to be determined in considerable detail. The structure of the outer few tens of kilometres varies considerably from place to place, and we will consider this shortly. The interior is much more uniform, seismic speeds changing with depth but not varying much with location.

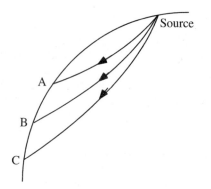

Figure 2.2 Because the speed of transmission of seismic waves increases with depth, the first vibrations detected at distant stations are not those that have travelled in a straight line, but those that have followed a curved track, shown here, taking them on a deeper yet faster route. The first vibrations would reach station A about 27 minutes after the event and station C only about 35 minutes after the event, even though C is about one and a half times further away.

What helps us to pin down the density distribution within the Earth is that at a depth of 2900 km the speed of seismic waves drops sharply. We can tell this because instead of continuing to follow an upward-curving track, the path of a vibration reaching this depth is deflected downward, as shown in Figure 2.3. At this depth the composition must change from rocky to something about twice as dense. This marks the outer limits of the Earth's **core**. The rocky material surrounding the core is known as the **mantle**.

The Earth's core

A remarkable thing about the core at this depth is that, despite the immense pressure, it is not solid. This has been demonstrated by studying the transmission of different kinds of seismic waves, which so far we have glossed over. There are two sorts of seismic wave that concern us when considering the Earth's interior: compressional waves and shearing waves. A compressional wave (or P-wave) is just like a sound-wave in air; it consists of alternate pulses of compression and expansion (rarefaction). A shearing wave (or S-wave) can be observed by shaking a jelly; it is an alternate side-to-side wobbling travelling through the body of the material. An important distinction between these two types of seismic waves is that whereas compressional

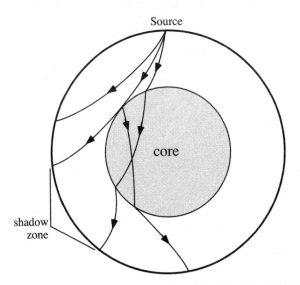

Figure 2.3 Paths of seismic waves deep within the Earth. The density of the core is about twice that of the overlying rocky mantle, and so seismic speed is halved. Waves are deflected sharply downward when they reach the core-mantle boundary, and there is a 'shadow zone' at a certain distance from the source within which seismic stations cannot pick up directly-transmitted seismic waves. In addition to earthquakes, underground nuclear explosions have proved to be useful sources of seismic energy.

waves can travel through anything, a liquid will not transmit shearing waves. This is because a liquid offers no resistance to shearing motion. The instruments that detect seismic waves (**seismometers**) can distinguish between the two types of waves, and from this we can tell that only P-waves are transmitted through the outer core, which must therefore be liquid.

That is not quite the end of the story, because we can identify an inner core whose limit is marked by a sudden increase in the speed of P-waves and within which we can tell (by indirect means) that S-waves are transmitted. Figure 2.4 is a plot showing the variation with depth of the speeds of P-waves and S-waves (which can be measured) and density (which we deduce from these measurements).

Now we have assembled the evidence needed to decide between the alternate models of density distribution given in Figure 2.1. We can see that the right-hand model (symmetrical distribution), rather than

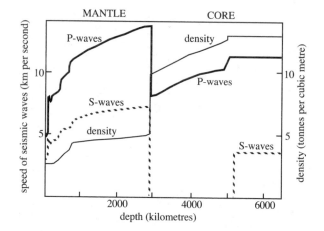

Figure 2.4 Speeds of compressional waves (P-waves) and shearing waves (S-waves) within the Earth, and inferred density.

the left-hand model (random distribution), is correct. In fact, we can go further than this, because we have enough to go on to propose what the dense stuff actually is. The solid inner core has the properties we would expect of solid iron mixed with a few per cent nickel. Its properties match that of other metals too, such as cobalt or titanium, but it seems logical that the Earth must be made of the same stuff as the Sun and meteorites, in which case an iron-rich model is the only reasonable interpretation. The liquid outer core has a density too low to be pure metal. About ten per cent of its mass must be composed of one or more relatively light elements. What these are cannot be proved, but oxygen, sulphur, carbon, hydrogen and potassium (a light metal) are the most likely.

Evidence from magnetism

There is one more factor bearing on the nature of the core that we should touch on here, and this is magnetism. The Earth has a magnetic field that looks like one produced by a giant bar magnet (Figure 2.5). Both common sense and the seismic evidence we have just investigated enable us to rule out a real bar magnet as the origin of this magnetic field. Instead, the field is regarded as the product of electrical currents generated by motion within the fluid outer core.

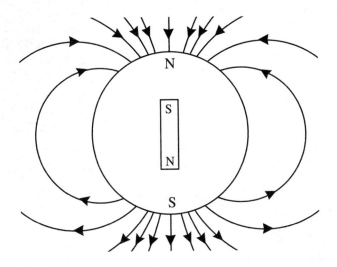

Figure 2.5 Cross-section through the Earth from pole to pole, showing the lines of magnetic field and the (imaginary) bar magnet at the Earth's centre.

The Earth's crust and mantle – compositional layers

Now we will turn to the outermost few tens of kilometres of the Earth, where seismic speeds are most complex. Junctions between one rock body and another are sometimes manifested by sharp changes in seismic speed, but on a global scale the effect is one of gradually increasing speed with depth. This is because of increasing rigidity with depth, brought about by pressure. P-wave speed increases gradually from about 2 km per second just below the surface to 6 or 7 km per second. Below this, a sharp jump to a P-wave speed of 8 km per second is recognised throughout the entire globe, at a depth of about 10 km below the ocean floor and at an average of 30 km below the continents. This boundary with more rigid rocks is known as the Mohorovičić discontinuity (after Andrija Mohorovičić, 1857–1936, the Croatian seismologist who first recognised it). It is usually called the **Moho** for short. Above the Moho are the rocks that belong to the Earth's **crust**, and below that we are into the mantle, which extends all the way to the core.

The crust is composed of slightly less dense rock than the mantle, and has a greater proportion of silicon, aluminium, calcium, sodium and potassium but less magnesium. It is easy to explain why the crust sits above the mantle; it is just the light stuff that has worked its way to the top. There are actually two distinct types of crust, continental crust (underlying virtually all the land surface and shallow seas) and oceanic crust (forming the floor of the deep oceans). The compositions of these two types of crust are shown alongside the mantle composition in Table 2.2.

element	oxide	Continental crust (granite-andesite)	Oceanic crust (basalt)	Mantle (peridotite)
silicon	SiO_2	62	49	45
titanium	TiO_2	0.8	1.4	0.2
aluminium	Al_2O_3	16	16	3.3
iron	Fe_2O_3	2.6	2.2	1.2
iron	FeO	3.9	7.2	6.7
magnesium	MgO	3.1	8.5	38.1
calcium	CaO	5.7	11.1	3.1
sodium	Na_2O	3.1	2.7	0.4
potassium	K_2O	2.9	0.26	0.03

Table 2.2 Average compositions of continental crust, oceanic crust, and the mantle. The names of the rock types most closely matching these compositions are shown. Elements are named in the first column, but compositions are expressed as the oxides of these elements whose chemical formula is shown in the second column. This is purely a convention and should not be thought of as indicating the chemical species actually present. Iron is listed twice, as Fe_2O_3 and FeO. SiO_2 is commonly referred to as 'silica'.

The thickness of the continental crust varies from about 25 km in thin, stretched regions to as much as 90 km below the highest mountain ranges. Oceanic crust is much thinner, ranging from about 6–11 km thick.

The wide variation in the thickness of continental crust is manifested not so much by mountains tens of kilometres high (Mt Everest reaches

only 8848 metres above sea level) as by enormous downward protru-
berances of the base of the crust, which we can detect as variations in
the depth of the Moho. This shows that mountains are not held up by
the strength of the material beneath them. Instead, crustal regions of
different thickness may be thought of as floating on the denser mantle. The situation is illustrated in Figure 2.6. Remember, though, that
the mantle cannot really be a liquid, because it transmits S-waves. It
just behaves like a liquid when considered over geological time.

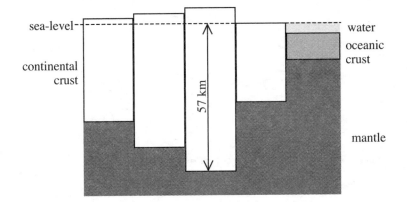

Figure 2.6 Cross-section showing crust of different thicknesses resting in equilibrium upon the mantle. In the situation shown, the weights of all five columns (crust
plus mantle, and including water in the right-hand column) measured from the
surface down to the base are equal. This demonstrates how mountains of modest
height (6 km above sea level in this example) are supported by buoyant roots.
Average densities of water, continental crust, oceanic crust and mantle are 1.0,
2.7, 3.0 and 3.3 tonnes per cubic metre respectively.

The concept of blocks of crust floating in equilibrium upon the mantle
is known as isostasy, and has been found to apply almost everywhere.
There are few examples known where high or low regions are held up
or pushed down by any force other than simple buoyancy (which is
really what is meant by isostasy). Buoyancy shows us why oceanic
crust is virtually never found on dry land. It is thinner and denser
than continental crust, and so it floats lower.

We can summarise what we have said so far about the Earth's
compositional layers by Figure 2.7. The compositional difference
between crust and mantle is relatively slight. The most abundant elements in both are silicon and oxygen. Any compound made of a chemical

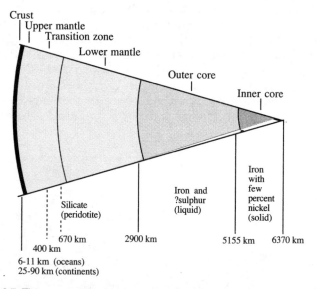

Figure 2.7 The compositional layers within the Earth. The lower mantle has a denser structure than the upper mantle, but is not thought to differ significantly in composition.

combination of just these two elements is known as **silica**. The rocky materials dominated by silicon and oxygen are therefore commonly referred to as **silicates**. Some changes in seismic speeds within the mantle have been recognised (see Figure 2.4), but these are thought to represent depths at which atoms within crystals become packed into denser, more rigid, structures, rather than being changes in chemical composition. Throughout, the mantle is thought to have a chemical composition similar to that of the rock type known as **peridotite**. The compositional distinction between the core and the mantle is much more fundamental than between the crust and the mantle. The core does not even consist of silicates, but is dominated by iron.

Another kind of layering – strong and weak

When we come to consider how the Earth behaves, it is not always chemical composition that is the most important factor. We also need to bear in mind how strong the material is. We have already met this concept when we considered the crust as blocks 'floating' in the mantle.

Mountain belts only need deep buoyant roots because the mantle does not have the strength to support them. Actually, the weakest zone in the mantle is not immediately below the crust, as you might think from a simple interpretation of Figure 2.6. Instead it begins at a depth of about a hundred kilometres. At this depth there is a slight drop in seismic speeds (discernable on Figure 2.4) which has been interpreted as due to the presence of a few per cent of molten material. Evidently, though, the material as a whole is still well and truly solid because S-waves can travel through it. This is known as the 'low speed layer' (sometimes the 'low velocity layer') and is weak enough to deform (over geological time periods) under the weight of mountain belts or ice sheets. For example, Scandinavia is still rising upwards at a rate of a few millimetres per year in response to unloading caused by the disappearance of a major ice cap that covered the region until about ten thousand years ago. It is flow in the weak zone in the mantle that is allowing this to happen.

You will discover in Chapter 4 that the outer shell of the Earth is broken into large slabs or 'plates' that are gliding sideways. This too would not be possible but for this weak zone, which lubricates their bases.

In terms of how the outer part of the Earth responds to forces, the crust and the uppermost part of the mantle constitute a single rigid unit known as the **lithosphere** (from 'lithos', the Greek word for rock). The lithosphere is rocky in terms of its mechanical properties, not just composition. In contrast, below the lithosphere the mantle is relatively weak, and the weak zone has been named the **asthenosphere** (using the Greek word for 'weak'). The relationship between crust, mantle, lithosphere and asthenosphere is summarised in Figure 2.8.

Figure 2.8 The most important layers of the Earth in terms of mechanical properties; the rigid lithosphere and the plastic asthenosphere.

The weakest part of the mantle lies in the few tens of kilometres immediately below the base of the lithosphere, and the term asthenosphere is often restricted just to this thin layer (coinciding with the low speed layer). However, once we get below the lithosphere there has been an important change in the properties of the rock that persists all the way to the core. This is that although solid so far as the transmission of seismic waves (which travel at many kilometres per second) is concerned, the deep mantle is not at rest. It is circulating slowly, at speeds of a few centimetres a year. This is usually described as 'solid-state convection'. Convection is what makes warm air rise and cold air sink, or water circulate in a saucepan (even before it boils). It is a way of transporting heat outward. Put simply, hot mantle rises upwards and transfers its heat to the base of the lithosphere. Having lost heat, this bit of the mantle becomes slightly denser, and sinks downwards to complete the loop (Figure 2.9). Most of the heat deposited at the base of the lithosphere trickles through to the surface by conduction, but some is carried higher by molten bodies of rock that can intrude high into the crust, or even reach the surface at volcanoes.

Figure 2.9 Possible arrangement of convection within the sub-lithospheric mantle. This model shows two layers of convection cells, which meet at the seismically-defined transition zone (see Figure 2.7).

The Earth's heat

Considering the nature of the mantle has brought us to the question of the Earth's heat. It is common knowledge that the interior of the Earth is hot, but most people probably think that this means that below a certain depth everything is molten. We have already seen that such a notion is misconceived, because seismic data prove conclusively that the mantle is virtually all solid.

The mantle *is* hot, but not as hot as you would think by extrapolating from surface measurements. Heat is escaping to the surface at a rate of about eight-hundreths of a watt per square metre (0.08 W m^{-2}), so you would need to collect all the heat escaping from an area greater than a 1000 square metres to power a 100 watt light bulb. Near the surface of continents, the average rate of temperature rise with depth (the geothermal gradient) is about 30 °C per kilometre. From this we could infer a temperature of about 900 °C at the base of 30 km thick crust and about 2700 °C at the base of 90 km thick crust, but we would be wrong to do so. Certainly temperatures over 1000 °C are not reached, because crustal rocks would melt at about this temperature.

The solution to this conundrum is that something like half the heat escaping to the surface of the continents is generated within the crust (we will see how shortly), and that only about half comes from below. This means that the geothermal gradient gets less as you go deeper.

If fact, below the lithosphere, temperature rises at a rate of less than half a degree per kilometre. One factor contributing to this low value is that, as we have already seen, this part of the mantle is convecting. Convection is much more effective than conduction at transporting heat. If convection could (magically) be stopped, then heat would build up within the mantle and its temperature would rise at a rate of a degree or so every million years until it melted. The solid-state convection within the mantle is actually what stops it becoming hot enough to melt.

The very effective convection in the liquid outer core keeps the temperature gradient even lower there, so that the temperature of the inner core is estimated at a surprisingly low value of approximately 4700 °C.

So where does the Earth's heat come from? An unknown proportion, perhaps around 30%, is heat that was trapped within the Earth as it formed. This primordial heat continues to leak out, but is not being generated today. The rest of the heat comes from sources that are continuing today, mostly from decay of radioactive elements.

Many elements have radioactive isotopes, but only three of these are important heat sources in the Earth today. These are potassium, uranium and thorium. Most potassium atoms are stable and non-radioactive. Each of these contains a total of 39 heavy particles (protons and neutrons) in its nucleus, and is described as potassium-39, or ^{39}K (using the chemical symbol for potassium, K). However, about one potassium atom in every ten thousand contains an extra neutron, making a total of 40 heavy particles. These atoms of potassium-40 (^{40}K) are unstable, and undergo radioactive decay. All atoms of uranium and thorium are unstable; thorium occurs as ^{232}Th, and uranium has two radioactive isotopes, ^{235}U and ^{238}U.

Although it is impossible to predict when any individual radioactive atom will decay, there is a measurable probability of it happening during a given time. When dealing with large numbers of nuclei, probability takes on a deadly certainty. We can therefore determine the time it takes for half the nuclei of a particular isotope to decay, which is known as the half-life. Because the number of radioactive nuclei halves over a fixed time, it never reaches zero. The half-life of ^{40}K is 4500 million years. This, coincidentally is roughly the age of the Earth, so it means that the abundance of ^{40}K in the Earth must now be virtually half what it was when the Earth formed. The half-life of ^{232}Th is over three times the age of the Earth, so about 80% of what the Earth started with is still there.

Of the four heat-producing isotopes, ^{40}K probably produces the greatest amount of heat today and ^{235}U the least. However, ^{235}U has the shortest half-life of the four, about 710 million years, which is about one-sixth the Earth's age. To determine the rate of heat production by ^{235}U when the earth was very young we must multiply the present rate by 2^6 or 64 times, from which we can deduce that it was the second most important heat-producing isotope during the first half billion years.

All three of these elements tend to be concentrated into the relatively light rocks of the continental crust, rather than the mantle, which explains why about half of the Earth's radiogenic (radioactively-produced) heat is generated in the crust. However, one of the great unknowns is whether potassium is abundant in the outer core, where it is a contender for the unknown light element. What is not in any doubt at all is that the Earth's radioactive power source must be running down, though at a rate that is too slow to concern us.

Radiometric dating and the age of the Earth

Apart from heat production, radioactive decay is important to the geologist because it is a physical process whose rate we know. By measuring the relative proportions of the surviving parent isotope and its decay products trapped within the same rock or mineral grain, the time since the rock or crystal was formed can be determined. For example, ^{40}K (the parent isotope) decays to ^{40}Ar (its decay product). As argon is not commonly trapped within minerals when they are formed, it is usually assumed that any argon found within a mineral is the product of decay of ^{40}K. Unfortunately argon can escape from within a crystal, so potassium-argon dates are sometimes unreliable.

The decay of uranium (to lead) offers an important dating technique, but because of its extremely long half-life ^{232}Th is not used much. There are other radioactive isotopes that are insignificant producers of heat but whose decay products enable **radiometric dating** (as it is called) to be performed. The most important of these is the decay of rubidium-87 (^{87}Rb) to strontium-87 (^{87}Sr). Radiometric dating techniques can be used to determine the time since a crystal grew from a molten rock or by recrystallisation of a rock without melting (a process known as metamorphism). Extensions of these techniques have enabled the age of the Earth, and of ordinary meteorites, to be determined at 4550 million years with a fairly high degree of confidence.

The origin of the Earth and the solar system

Now that we have completed our brief survey of the Earth's interior and gross composition, it is time to describe how scientists think the Earth came into being. This takes us back to the birth of the Sun, which grew by the collapse of a slowly rotating interstellar cloud of gas (mostly hydrogen) and dust. This cloud is referred to as the solar nebula. As it contracted, the solar nebula span faster and faster. The material that was not drawn into the central point, where the Sun was forming, became concentrated in a disc around the Sun. It was within this disc that the planets grew. At first the cloud would have been very hot, because of the gravitational energy converted to heat by the contraction. As it cooled, things began to condense out of the gas as tiny solid grains. The first grains to condense would have been made of substances able to form at temperatures above a thousand degrees. These would have included nickel-iron metal and some of the silicate minerals. As the temperature declined to a couple of hundred degrees centigrade the minerals growing could begin to trap water within their structure, but ice crystals would not begin to grow until the temperature reached –90 °C.

The tiny grains that formed within the solar nebula would tend to stick together whenever they happened to come into contact, and so progressively larger chunks would collect. Once the process had begun, it may have taken as little as a few thousand years to form centimetre-sized pieces. However, throughout the sequence of events the timescales are poorly understood. After about 100 thousand years, the biggest blocks had probably grown to about ten kilometres across and are dignified with the name 'planetesimals'. These were now big enough to make their gravitational influence felt, and their growth would now proceed at an accelerating rate until after about a further 50 thousand years most of the planetesimals had collided and accreted into a few dozen bodies a few thousand kilometres across, now known as 'planetary embryos'.

Collisions between these surviving bodies would have been violent, sometimes shattering both bodies, more often pasting the debris of the smaller body across the face of the larger one and liberating sufficient heat to cause melting to a very great depth. After about 100 million years all the planetary embryos that were going to collide had proba-bly done so, leaving the solar system with four inner rocky planets (see Table 2.1), four giant planets (Jupiter, Saturn, Uranus and Neptune) further out from the Sun where it had become cold enough for large quantities of water, methane, ammonia and similar volatile sub-stances to condense, and relatively tiny icy objects (the planet Pluto being one of the largest) in the outer fringes where the density of material was too sparse to collect into large bodies.

The Earth probably acquired the Moon during these latter stages. Although still controversial, it is likely that the Moon is formed from the debris of the last planetary embryo collision to affect the Earth. In this case, some fragments of the impacting body bounced back from the Earth and collected together in orbit around the Earth.

The irregularly shaped rocky and iron-rich bodies known as the aster-oids, which are up to a few hundred kilometres across and found most-ly between the orbits of Mars and Jupiter, were once thought to be the remains of a planet that was broken up by a giant collision. However, it now seems more likely that they are surviving planetesimals that never stuck together when they hit each other, because their orbits were stirred up by their proximity to Jupiter so that they hit each other too hard to allow accretion. Most meteorites, which are chunks of rock or nickel-iron that fall from the sky, are thought to be made of the same stuff as the asteroids.

The story as outlined above may sound rather glib. It is important to realise that we are on the threshold of the realm of legend here. Most scientists would accept this story in outline, but there are many important details still in dispute. For example, the temperature history of the solar nebula is uncertain. It could have remained too hot for much rocky material to have condensed until substantial sized bodies of nickel-iron had formed. In this case, the rock would have collected around these pre-grown cores and planetary bodies would develop with an in-built layered structure. On the other hand, temperature may have fallen swiftly enough for chunks of a wide range of compositions to have been around at the same time. Planetary bodies would then have grown with a thoroughly mixed composition, and internal layering would then have to be generated by some later process.

It is actually not too difficult to envisage how compositional layering could develop from initially well-mixed material, especially if we choose to believe that a planet like the Earth grew by collisions between planetary embryos of similar size. As noted above, these impacts would have been violent enough to cause global melting, and this would have provided the opportunity for dense phases, such as iron, to sink and form the core, and lighter materials, such as the silica-rich minerals characteristic of the crust and mantle, to rise upwards. However, a variant of the standard theory omits the planetary embryo stage, and has each planet developing from the largest planetisimal in its vicinity, which grew by accreting other planetesimals that were always much smaller than itself. In this scenario, the heat released by impact energy may not have been enough to cause global melting, in which case differentiation into a compositionally layered structure would have to depend on internal sources of heat, such as radioactive decay.

We will put speculation aside now, and turn to much firmer ground (metaphorically speaking!) and look at what happens in an earthquake.

3

EARTHQUAKES

We have already seen how useful earthquakes are for probing the interior of the Earth. This is because the seismic waves generated by earthquakes can travel right through the Earth. However, earthquakes themselves do not happen deep down, in fact they are virtually confined to the lithosphere. It is their shallowness that makes the larger ones so devastating. In this chapter we will concentrate on describing earthquakes. Their fundamental causes will become apparent in the next.

How earthquakes happen

There are many fault-lines in the Earth's crust. The San Andreas fault in California is probably the most famous. A fault is where blocks of crust on either side are moving relative to one another. The typical average rate is around a millimetre per year. If this movement happened gradually it would pose few problems for people living nearby. Unfortunately rocks do not behave that way. Instead, they tend to stick. Strain builds up for decades or centuries until it reaches a critical level, and then everything gives at once. The principle is shown in Figure 3.1. Something similar to what goes on at the break-point of an earthquake happens if you stack two house bricks on top of each other. If you hold the lower one, and gradually tilt it you will find that the upper one does not move at first. As the angle of tilt increases, the force tending to make the top brick slide gradually increases until it is great enough to overcome friction, and suddenly it will slide off.

Figure 3.1 Plan view showing a road built across a fault (dashed line). Strain builds up gradually, (a)–(c), as the edges of blocks on either side of the fault are bent elastically, because friction is too great to allow them to slide past each other. Eventually (d) the fault gives way and rapid slip occurs, liberating a lot of seismic energy.

Once a fault has given way at one point, slip movement may occur along its whole length, which may be hundreds or thousands of kilo-metres, though movement is usually restricted to a much shorter por-tion of the whole fault. The strongest seismic waves are generated at the initial break-point. The closer to the break-point, the greater the energy and the greater the potential for destruction, especially if the break-point is near the surface. Technically, the break-point is called the earthquake **focus**, and the point on the surface directly above it is referred to by that much-abused term **epicentre**. Slip further along the fault and readjustments close to the focus usually cause a series of smaller aftershocks, which continue for days (even years in extreme cases) after the initial earthquake.

Most of the damage caused by an earthquake is not done by the P-waves and S-waves that we considered in the previous chapter. These waves travel through the body of the Earth, and their energy is spread over a rapidly increasing volume as they propagate. It is other waves of the sort that travel along the surface of the ground that do most of the harm. These include up-and-down waves (like waves on the ocean) and strong side-to-side shaking waves. The P- and S- body waves travel faster than the surface waves, and, if felt, can give a few seconds (minutes, if further away) warning of the arrival of the more damaging surface waves.

Measuring earthquakes

When an earthquake makes the news, its size is usually described by a number on the Richter scale. This scale (of which there are several variants) is an attempt to document how strong the seismic waves must have been at the focus. This can be estimated using the signal detected by a single seismometer at a known distance from the focus, or, better, by comparing the signals detected at many seismometers at a variety of distances. The Richter scale is numerical, enabling comparative data to be compiled on earthquakes of all sizes from around the globe. For each one magnitude increase on the scale, the motion (or amplitude) of the seismic waves at the focus increases tenfold and the energy released increases roughly thirtyfold. The most devastating earthquakes tend to be of magnitude 6 and above. Fortunately there is on average only one earthquake bigger than magnitude 8.0 each year. Table 3.1 summarises the occurrence of earthquakes on the Richter scale.

Magnitude range	Number per year	Example
8.0 and above	1	San Francisco 1906 M8.3 (500 deaths) Tokyo 1923 M8.2 (143 000 deaths) Kurile Islands 1994 M8.2 (12 deaths)
7.0–7.9	18	Tangshan, China 1976 M7.6 (250 000 deaths) Mexico City 1985 M7.9 (30 000 deaths) Loma Prieta (San Francisco) 1989 M7.1 (61 deaths) N Iran 1990 M7.6 (40 000 deaths) East Cape Town, North Island, New Zealand 1995 M7.5 (no deaths) Kobe, Japan 1995 M7.2 (5000 deaths)
6.0–6.9	108	Avellino, Naples, Italy 1980 M6.8 (8000 deaths) Spikak, Armenia 1988 M6.8 (25 000 deaths) Northridge (Los Angeles) 1989 M6.7 (61 deaths)
5.0–5.9	800	Cook Strait, New Zealand 1995 M5.3 (no deaths)
4.0–4.9	6 200	Coalinga, California 1983 M4.3 (no deaths); Shropshire, England 1990 M4.5 (no deaths)
3.0–3.9	49 000	
2.0–2.9	300 000	
<2.0	millions	

Table 3.1 Numbers of earthquakes per year in each magnitude range of the Richter scale. Some well known, or typical earthquakes of each magnitude are listed.

A disadvantage of the Richter scale is that it does not show how much damage a particular earthquake is capable of. For example, deep focus earthquakes are not very effective at generating surface waves, so they tend not to cause much damage. Furthermore, what effect an earthquake has depends on when and where it occurs. A magnitude 8 earthquake in the middle of nowhere will probably cause much less damage than a magnitude 6 earthquake with its epicentre in a major city, especially if it happens during the rush hour. The effect of an earthquake also depends on the nature of the rock near the surface. In the Mexico City earthquake of 1985 the soft sandy subsoil quivered so vigorously that buildings sank deep into it, causing many deaths. A similar earthquake below a city built on hard rock would probably have had much less effect.

An alternative to the Richter scale is the Mercalli scale. This ranks earthquakes according to the intensity of their effects, and can also be used to map the extent to which concentric areas are affected by a single earthquake. A version of it is shown in Table 3.2.

Mercalli intensity	Description
XII	Damage total. Lines of sight and level distorted. Objects thrown in the air.
XI	Few if any masonry structures remain standing. Bridges destroyed. Rails bent greatly.
X	Some well-built wooden structures destroyed; most masonry and frame structures destroyed with foundations. Rails bent.
IX	Damage considerable in specially designed structures; well-designed frame structures thrown out of plumb. Damage great in substantial buildings, with partial collapse. Buildings shifted off foundations.
VIII	Damage slight in specially designed structures; considerable damage in ordinary substantial buildings, with partial collapse. Damage great in poorly-built structures. Fall of chimneys, factory stacks, columns, monuments, walls. Heavy furniture overturned.
VII	Damage negligible in buildings of good design and construction; slight to moderate damage in well-built ordinary structures; considerable damage in poorly-built or badly designed structures; some chimneys broken. Difficult to stand upright. Noticed by vehicle drivers.

Mercalli intensity	Description
VI	Felt by all, many frightened. Some heavy furniture moved; a few instances of fallen plaster. Damage slight.
V	Felt by everyone; many awakened. Some dishes and windows broken. Unstable objects overturned. Pendulum clocks may stop.
IV	Felt by many indoors, outdoors by few. At night, some awakened. Dishes, windows, doors disturbed; walls make cracking sound. Sensation like heavy truck striking building. Standing cars rock.
III	Felt quite noticeably by people indoors, especially on upper floors. Many people do not recognise it as an earthquake. Hanging objects swing. Feels like passing traffic.
II	Felt only by a few people at rest, especially on upper floors.
I	Generally detected by instrument only.

Table 3.2 The Modified Mercalli scale of perceived earthquake intensity. The more intense earthquakes are categorised by structural damage, less intense earthquakes by effects perceived by people on the spot.

Some of the descriptions on the Mercalli scale are universally applicable, such as whether the earthquake can be felt by people who are walking about, or only by those lying down. However, the amount of damage to buildings depends very much on the nature of the construction. A well-designed structure can withstand even a very major earthquake without collapsing, even though it may be so badly damaged that the only safe thing to do afterwards is to demolish it. In many earthquake-prone regions, building codes are enforced to ensure that structures are relatively collapse-proof.

Protecting structures from earthquake damage

The are many ways to make buildings, elevated highways, and other structures earthquake resistant. It is not simply a matter of making everything stronger. It is more important to ensure that if a structure is rocked by an earthquake then the whole of it should sway in the same direction together (Figure 3.2).

Figure 3.2 Two series of cross-sections to show possible response of a simple structure (two walls and a roof) to the shaking caused by an earthquake. In (a)–(d) both walls sway in the same direction simultaneously. The structure survives with relatively minor damage. In (e)–(h) the walls sway in opposite directions, with catastrophic results. At the very least, the roof collapses even if, as here, the walls manage to stay upright. In inadequately designed multistory buildings, several floors have been known to collapse on top of one another, squashing anyone unfortunate enough to have been inside.

As well as tying walls together so that they sway in unison it is important to ensure that a building can sustain any twisting motion engendered by the earthquake. Other things that can help are to dig extra-deep foundations, so that the structure rests on solid bedrock where the amplitude of the vibrations is likely to be less, or at the least to make sure that the foundations rest on the same material all round so any subsidence happens evenly on all sides.

Another factor is the natural frequency at which a building tends to shake. Taller buildings sway more slowly than low-rise buildings, and there is a danger that high-rise buildings will be 'in tune' with the shaking of the ground. This had an unfortunate consequence in the major earthquake that struck Mexico City in 1985. In the most heavily damaged area of the city, about 70% of buildings between 10 and 20 stories high were damaged because the low-frequency shaking of the ground became amplified within the building by resonant vibrations of their structure. In contrast, fewer than 20% of buildings less than five or more than 20 stories high were damaged.

Searching for survivors after an earthquake is a risky business. Buildings that are apparently unscathed may actually be dangerously near to collapse, which could be triggered by an aftershock any minute. There may also be a risk of explosions from ruptured gas pipes.

Fire is actually one of the biggest hazards in the aftermath of a major earthquake. Most of the death and destruction associated with the San Francisco earthquake of 1906 was not a result of collapsing buildings but of fires, caused by toppled lamps and the like. The majority of the buildings survived the quake, but were razed to the ground by the subsequent fires.

Many lessons have been learned about how to build earthquake-resistant structures. However, even in the wealthier at-risk areas large numbers of old and less-resistant structures remain, as demonstrated by some of the damage caused by recent earthquakes in California (Figure 3.3) and Japan.

Figure 3.3 This building in San Francisco was wrecked when its lower floors gave way during the 1989 Loma Prieta earthquake.

Earthquake preparedness

In earthquake-prone regions most people are well aware of the risks they face. Instructions on how to be prepared, and what to do in the event of an earthquake are widely available. The following is an extract from the advice pages of the Los Angeles telephone directory.

During an Earthquake

1 If you are indoors, **DUCK** or drop down to the floor. Take **COVER** under a sturdy desk, table or other furniture. **HOLD** on to it and be prepared to move with it. Hold the position until the ground stops shaking and it is safe to move. Stay clear of windows, fireplaces, and heavy furniture or appliances. Don't rush outside. You may be injured by falling glass or building parts.

DO **NOT** try using the stairs or elevators while the building is shaking or while there is danger of being hit by falling glass or debris.

2 If you are outside, get into the **OPEN**, away from buildings and power lines.

3 If you are driving – **STOP** if it is safe – but stay inside. DO **NOT** stop on or under a bridge, overpass or tunnel. Move your car as far out of the normal traffic pattern as possible. DO **NOT** stop under trees, lightposts, electrical power lines or signs.

4 If you are in a mountainous area, be alert for falling rock and other debris that could be loosened by the quake.

5 In a crowded public place, do **NOT** rush for the exits. Stay calm and encourage others to do so.

Even when there is no structural damage, the shaking that occurs during an earthquake can cause breakage and injury. Simple precautions include not hanging pictures, especially glass-fronted ones, over the head of one's bed, and securing china ornaments in place with 'earthquake wax'.

Prediction and prevention

The business of predicting earthquakes is not simple. Usually, the first rupture is the biggest, and the aftershocks in the following weeks are of decreasing magnitude. If, instead, earthquakes usually built up towards a crescendo it would be more obvious when a big earthquake was due! As it is, one of the most effective ways of predicting earthquakes is to map out areas where earthquakes have *not* been happening lately (Figure 3.4). If a length of fault can be identified with no recent break-points, but the lengths of fault to either side have many

Figure 3.4 Map of a fault and the surrounding area, showing earthquake epicentres over a thirty-year period. The central part of the fault has not ruptured recently, and the prospects of a major earthquake affecting the nearby city in the next few years look pretty grim.

recent epicentres, then it is probable that a large amount of strain has built up in the seismically quiet bit in the middle. This strain is likely to be released soon by a major rupture, as when an elastic band, stretched to breaking point, suddenly snaps. The trouble is that the method of pinning down exactly when this will happen has so far proved elusive.

There is not much that can be done to prevent earthquakes happening. One suggestion has been to pump water into fault-planes in areas where strain build up is indicated by lack of recent seismic activity. The hope is that the water will act as a lubricant, allowing the crustal blocks to begin to slip past each other gently instead of violently. However, such a course of action is risky. Imagine the chances of re-election for a local government officer on whose orders this was done if, in order to forestall a magnitude 8 earthquake that was likely sometime in the coming decade, the lubricated fault moved in such a fashion as to cause a series of magnitude 5 earthquakes. Even if only minor damage occurred, the body responsible for the fault lubrication programme would get the blame, and the lawyers would have a field-day trying to establish financial liability!

'Tidal waves'

One other aspect of earthquakes deserves to be mentioned, and this is the effect of earthquakes in the ocean. If an epicentre is below the sea-floor, the sudden displacement of rock can create a wave travelling through the ocean at a speed of a couple of hundred metres per second. In deep water, such a wave may be scarcely noticeable, being only about a metre high. However, upon reaching shallow, coastal water the wave slows down and gets both higher and steeper with the result that it may break upon the shore with tremendous force, sweeping buildings (and people) away even several tens of metres above sea level. Popularly (but incorrectly) called a 'tidal wave', this phenomenon is more correctly referred to by its Japanese name of **tsunami**. A tsunami can also be triggered by landslides or undersea volcanic eruptions.

The wide-ranging effect of tsunamis is demonstrated by the case of the Hawaiian Islands, whose north shores face the earthquake-prone belt of Alaska and the Aleutian Islands, over 3000 km away. The largest historic tsunami to hit Hawaii happened in 1946, and was triggered by an earthquake in the Aleutians. This wave broke on the shore and reached up to 17 m above normal sea level, sweeping away entire villages and taking the lives of 159 people. Today, tsunami-warning yellow speakers mounted on telephone poles are a familiar sight in parts of the Hawaiian islands. It takes a tsunami several hours to travel from the Aleutians to Hawaii, but there is much less warning time in the case of tsunamis generated locally.

4
PLATE TECTONICS

We have so far discussed earthquakes without paying much attention to where in the world they tend to occur. Certain parts of the globe suffer more earthquakes than others, and you will have experienced at least a few minor tremors if you live in a high-risk area. A glance at a map of earthquake distribution, such as Figure 4.1, shows that they tend to occur in belts. The most famous of these stretches all the way up the west coast of the Americas, though the Aleutian and Kurile islands and southwards round the western side of the Pacific into New Zealand, so that the belt almost encircles the Pacific basin.

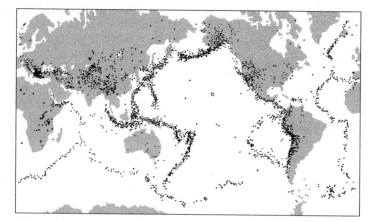

Figure 4.1 Map showing the distribution of earthquake epicentres over a typical six-year period.

This belt is also where many of the most active on-land volcanoes occur, and has been called the Pacific Ring of Fire.

The coincidence of earthquakes and volcanic zones is no accident, and the relationship between between these two quite different phenomena will become clear in this chapter and the next. Earthquakes also occur in a broad belt running through the Mediterranean and eastwards into central Asia (where there are very few volcanoes), and in some much narrower zones snaking their way through the oceans, notably a zone running from north to south through the Atlantic.

Moving plates

Earthquake zones are a manifestation of a phenomenon known as **plate tectonics**, though each of the three zones mentioned above fits into the picture in a different way. Plate tectonics is the term used to describe the manner in which the Earth's lithosphere moves around. As established in Chapter 2, the relative weakness of part of the upper mantle allows the lithosphere to slide around. However, the lithosphere is not a single unit. It is broken into a series of rigid plates, usually described as seven major ones and half a dozen minor ones (these are named on Figure 4.9, on page 42). Most plates carry areas of both continental and oceanic crust. The Pacific plate is a notable exception in having no large areas of continent.

Each plate is in contact with its neighbours on all sides, but the plates are moving relative to one another. It is important to realise that there are no gaps between these plates, so there are no chasms open clear down to the asthenosphere. To see how plates are able to move around without any gaps appearing, we will look at processes at the margins of plates.

Plates colliding

First, what happens when plates are moving towards each other? This situation is typified by Japan, shown in cross section in Figure 4.2. Japan is a piece of continental crust on the eastern edge of a major plate. The floor of the Pacific ocean belongs to a different plate, which is moving towards Japan at a rate of about 10 cm per year. Where the two plates meet, one is being thrust down below the other, a process described as **subduction**. Because oceanic crust is denser

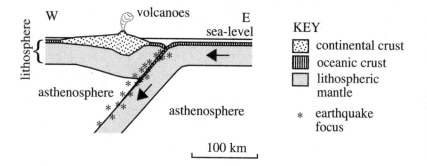

Figure 4.2 Cross-section through Japan; see text for explanation.

than continental crust, when two plates meet it is almost invariably the oceanic plate that goes under, as in this example. Under Japan, the Pacific Plate forms a slab descending at an angle of about 45°, but examples of much steeper and much shallower subduction are found in comparable situations elsewhere.

As one plate slides over another, the front of the over-riding plate is compressed, and the rocks there may become buckled. At the plane of movement between the two plates, motion is far from uniform, progressing in the stick-slip fashion that we considered in the previous chapter. This gives rise to earthquakes, and it was by plotting the depths of earthquake foci that subduction zones were first recognised for what they are. As you can see in Figure 4.2, earthquakes become deeper from east to west under Japan.

Inclined zones of earthquakes can be traced to about 700 km, but no deeper. This is because, as a plate descends, heat from the surrounding asthenosphere warms it, so it eventually ceases to be recognisable. The first part of the subducting slab to lose its identity is its crust, because crustal rocks melt at a lower temperature than the temperature of the sub-lithospheric mantle that they encounter. The melted material rises upwards to feed volcanoes, which tend to occur in a belt about 70 km above the subduction zone. We will look at such volcanoes in more detail in the next chapter.

The top end of a subduction zone is marked by a trench on the sea floor. Off northern Japan the trench is a fairly typical 8 km deep, but further south the trench reaches over 11 km, which is the greatest known depth in the world's oceans.

When two plates meet at a subduction zone one of them is destroyed, so this setting is described as a **destructive plate boundary**. However, it is only the oceanic part of a plate that can be destroyed in this way. Let's consider what happens when *both* plates contain continents (Figure 4.3). At first, subduction proceeds as normal, while the oceanic

Figure 4.3 Time series of cross-sections to show a continent–continent collision. See text for explanation.

part of one plate descends below the other. However, eventually the continental part of the subducting plate reaches the subduction zone. Continental crust is thicker and more buoyant than oceanic crust, and this prevents the plate from continuing to subduct. The edges of both continents are buckled, but one will eventually get thrust over the other. Soon afterwards, the subduction zone jams. The oceanic part of the descending slab breaks free, leaving the two plates joined together above. Near this suture the crust may be double its normal thickness, and this is where the highest mountains are to be found.

A recent example of such a continent–continent collision began about 30 million years ago when India collided with Asia. We see the aftermath of this event in the high elevation of the Tibetan plateau, fringed to the south by the Himalayas. Another consequence of this collision, and others that accompanied it, is the diffuse belt of earthquakes running eastwards from the Mediterranean through central Asia. As for the vanished ocean, the only remains are a few slivers of oceanic crust and upper mantle caught up in the suture zone, known as **ophiolites**.

Plates moving apart

So much for what happens where two plates are moving towards each other. This cannot be the only sort of interaction between plates, because if it were the Earth's surface would have to be shrinking all the time. In fact, we now know that new ocean floor is being created at a rate sufficient to compensate for the loss of oceanic parts of plates at destructive plate boundaries. Unsuprisingly, these sites are known as **constructive plate boundaries**.

What goes on here is summarised in Figure 4.4. As two oceanic plates are drawn apart, in a process referred to as sea-floor spreading, the underlying asthenosphere wells upwards to avoid any gaps appearing. As the upwelling asthenospheric mantle approaches the surface it cools, and becomes part of the lithosphere belonging to the plates on either side of the boundary. This new lithosphere is still relatively warm, which makes it slightly less dense than the older, colder lithosphere further from the boundary, so these boundaries are marked by ridges. Typically, the crest of such a ridge lies at a depth of 2–3 km below sea level, whereas the expanse of ocean floor to either side is at an average depth of 4–5 km. There are earthquakes associated with

Figure 4.4 Cross-section through a constructive plate boundary, where upwelling asthenosphere accretes to the diverging edges of two lithospheric plates. The oceanic crust is a result of partial melting of the upwelling asthenospheric mantle (see text for explanation).

the rifting and subsidence of oceanic lithosphere near constructive plate boundaries, but these are all at shallow depths, giving a seismic belt entirely different in character to a subduction zone.

In addition, a small percentage of the upwelling mantle melts. This is not because there is a heat-source here, it is a consequence of the drop in pressure, and is known as decompression melting. When mantle of peridotite composition (about 45% silica, SiO_2) begins to melt, the molten rock (or **magma**) produced has a slightly higher silica content. This is what gives rise to oceanic crust (Table 2.2) where the composition is, on average, about 49% silica. This composition is described as 'basic' or 'basaltic', because it matches the composition of the rock-type known as **basalt**. The residual mantle left behind has a reduced silica content to compensate for the enrichment of silica in the crust, but because the volume of melt produced is very much smaller than the volume of mantle contributing to the melt, the chemical change in the mantle is slight.

The ability of rock to melt in this way, so that a small volume of melt enriched in silica (and other components) is produced, is a very important process in geology, and is known as **partial melting**. The whole of the oceanic crust has been produced by partial melting of the mantle. We will look in more detail at how the oceanic crust is constructed, and in particular at the sub-sea volcanism associated with constructive plate boundaries, in the next chapter.

Magnetic stripes on the ocean floor

One of the most compelling lines of evidence in favour of sea-floor spreading is provided by magnetism. Rocks that have cooled down from a melt inherit their magnetisation from the Earth's magnetic field at the time they cool. Such rocks forming today have their magnetisation aligned with the present day field. However, the direction of this field has flipped many times in the past, and rocks that formed when the field was reversed are reversely magnetised. It is not known why the field reverses, but it provides another argument, if one were needed, for discounting the idea of a giant bar magnet within the Earth.

Studies of the magnetisation of the sea-floor carried on during the 1950s by towing sensitive instruments behind research vessels began to reveal a curious stripey pattern of magnetisation on the sea-floor.

Half the stripes are magnetised in the direction of the present field, but half are magnetised in the reverse direction. This was elegantly explained in 1963 by two British geoscientists, Fred Vine and Drummond Matthews, who suggested the explanation of a mid-ocean 'tape recorder' (Figure 4.5). As new ocean floor is added at a constructive plate boundary it is magnetised according to the direction of the Earth's magnetic field, and it retains this magnetisation as it moves away. Reversely magnetised stripes represent ocean floor formed at times when the Earth's field was in the reverse direction.

Figure 4.5 (a) The mid-ocean tape recorder. Normal (black) and reverse (white) magnetic stripes on the ocean floor show the polarity of the Earth's magnetic field at the time when each piece of ocean floor was added to the edge of the plates at a constructive plate margin. (b) The magnetic polarity time-scale.

The dates of reversals in the Earth's magnetic field extending back for a couple of hundred million years have been determined by studying the magnetisation of volcanic rocks on land, whose age has been established radiometrically. As a result, the age of the ocean floor at any point can be told just by counting back the number of magnetic stripes from the constructive plate boundary, or by recognising a distinctive pattern of reversals of different lengths, without the need to collect a sample.

The break-up of a continent

If a linear belt of asthenospheric upwelling is initiated below a continent instead of an ocean, the continental crust at first becomes stretched thin, then it ruptures, a constructive plate margin develops in between, and new *oceanic* crust is formed. This is illustrated in Figure 4.6, and something of the sort seems to have been responsible for the origin of the Atlantic Ocean. The apparent fit of South America and Africa together if you imagine the Atlantic closed (and not just the fit of their coastlines, but of more fundamental geological structures too) has long been recognised and given rise to theories of moving continents, described as **continental drift** before the discovery of plate tectonics in the 1960s provided the explanation.

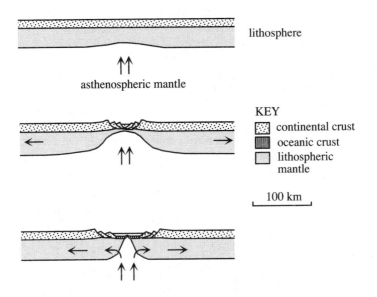

Figure 4.6 Time series of cross-sections to show the splitting of a continent to form a new ocean. Once sea-floor spreading has begun, as in (c), it continues at a rate typically between 1 and 20 cm per year. Subsidence along some of the faults that were initiated during the initial rifting event may continue sporadically at the rifted edges of the continents for tens of millions of years. This is the biggest cause of earthquakes not at plate boundaries.

The life-cycle of an ocean

The rates of ocean floor creation, though varying from place to place, are always approximately equal on either side of a constructive plate boundary. In consequence, the ridge marking this site in the Atlantic Ocean is mid-way between the two shores. Constructive plate boundaries in general are therefore often called 'mid-ocean ridges'. This is a misnomer, because eventually the oceanic part of one or other of the plates becomes detached from the continental part and begins to subduct beneath it (Figure 4.7), as has happened in the Pacific and Indian Oceans. Whether or not the ocean continues to get wider depends on the relative speeds of subduction and sea-floor spreading, but whatever happens the constructive plate boundary cannot stay in the middle.

Figure 4.7 shows what will almost certainly be the eventual fate of the Atlantic Ocean. If we imagine this as a west–east cross-section then it implies subduction (with its attendant earthquakes and volcanoes) beneath New York. However, it could equally well happen on the European side instead. Eventually, the constructive plate boundary could get drawn into the subduction zone, and once this has happened the ocean must grow narrower until, as in Figure 4.3, there is a continent–continent collision.

It is thought that the cycle of continental splitting, ocean opening, ocean closing, and continental collision takes on average 400–500 million years. You may have noticed when reading the above that whereas ocean floor is continually created and destroyed, there has been no

Figure 4.7 Time series of cross-section to show the evolution of the ocean, formed in the way shown in Figure 4.6. The scale is reduced here, and the crust has not been distinguished. The stage reached by the Atlantic Ocean is shown by (a), though it is wider than shown here. In stage (b) the left-hand plate has split into two, and the oceanic part is being subducted beneath the continental part. The Pacific Ocean is in stage (c).

mention of formation of continental crust. Continental crust is split and re-assembled in ever-changing configurations by the action of plate tectonics, but it is never destroyed, and nor does it grow except by the addition of small volumes resulting from volcanic activity above subduction zones. We will postpone consideration of the origin of continental crust until Chapter 13, except to say that most is thought to go back over at least 2000 million years, although much of it has been deformed many times since then, and that some traces date back nearly 4000 million years. In contrast, because it is recycled, the oldest known oceanic crust is a mere 190 million years old.

Plates sliding past each other

Not all boundaries between plates are sites of either convergence or separation. Some mark places where plates are sliding past each other. These are known as conservative plate margins. Figure 4.8 shows a typical example within an ocean, where a constructive plate boundary is offset by a fault, known as a **'transform fault'**. A transform fault may offset the ridge by anything from ten to a couple of hundred kilometers. It is marked by a narrow zone of shallow earthquakes. The transform fault is the bit between the two offset ridge segments; its apparent continuation beyond this lies within a single plate, and is not a plate boundary. This is known as a 'fracture zone' and is often manifested as a sharp change in depth because of the different ages of ocean floor on either side of it, but it has far fewer earthquakes because there is no sideways slip between the rock on either side.

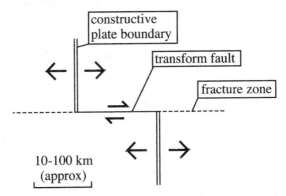

Figure 4.8 Map view showing how a transform fault offsets a constructive plate boundary.

Conservative plate boundaries can run through continents too. The most celebrated of these, of which the San Andreas fault is a part, runs through California. Here, the part of California on the south west of the fault is attached to the Pacific plate, and is moving north-west-wards at an average rate of 1 cm per year relative to the rest of north America. It is stick-slip motion along this fault system that causes most of the earthquakes in the San Francisco–Los Angeles region.

The global picture

We are now in a position to view the big picture of how the plates and the boundaries between them fit together on a global scale. This is shown in Figure 4.9. There are essentially two settings in which destructive plate boundaries can occur. The first is where one oceanic plate descends below another. When this happens, volcanism on the over-riding plate leads to the construction of a series of volcanoes, which is the origin of the volcanic 'island arcs' in the western Pacific. On the other hand, when subduction occurs beneath a continent, as where the Nazca Plate (south-east Pacific) subducts beneath South America, the volcanoes grow on pre-existing continental crust, giving

Key: Plate Margins   ⊥⊥⊥⊥ destructive   ——— conservative   =⊏ constructive
     ◄10 direction of motion (speed in mm per year)

Figure 4.9 Global map on the same base as Figure 4.1, showing plate boundaries and the rates of plate motion, relative to Africa (virtually stationary). Recent continental collision zones are shown here as destructive plate boundaries.

rise to an Andean-type volcanic mountain range. The case of Japan, with which we began, is something in between. Part of Japan is old continental material rifted away from the eastern edge of the Asian continent, but much of it is young island arc material.

A potentially confusing feature of any map such as Figure 4.9, is that it attempts to portray on a flat sheet of paper something that is really happening on the surface of a sphere. Tectonics on a sphere means that rates of relative plate motion must vary from place to place along plate boundaries. For example, the constructive plate boundary between the Pacific plate and the Antarctic plate is spreading at a rate of less than 6 cm a year between Antarctica and New Zealand, but the rate gradually doubles as the boundary is traced over the next 1000 km north-eastwards. Another is that the relationships between individual plates are continually readjusting to prevent any gaps opening up between plates. This effect is particularly noticeable in trying to trace through time the positions of triple junctions where three plate boundaries meet.

So far we have kept the discussion simple, by assuming that the relative motions between plates are parallel. Although this is commonly a good approximation, especially for constructive plate boundaries, it does not always hold. There are plenty of examples of where collisions between continents have occurred obliquely. Indeed much of the western coast of north America from California to Alaska is thought to consist of pieces of continent described as 'exotic terranes', that slid in obliquely from the south at various times during the past 200 million years. To take another example, conservative plate boundaries within the continents often have kinks in them. In the example shown in Figure 4.10, the relative motion as seen looking from one side of the fault to the other is from left to right. Overall, the fault runs parallel

Figure 4.10 Map view, to show how, where a conservative plate boundary is not exactly straight, zones of compression and tension develop along it.

with the direction of relative motion, but where the fault line bends to the left there is a zone of compression (manifested by buckling of rocks and thrusting on one unit over another) and where it bends to the right there is a zone of extension where the crust is stretched thin. The San Andreas fault system is of this type.

What makes it happen?

In all this discussion of plate tectonics, we have not considered what makes it happen. This is still controversial, but most geologists would agree on two things, at least. The first is that, attractive as the idea might seem, plate motions are not the direct surface expressions of mantle convection of the kind illustrated in Figure 2.9. The plates seem to be moving around more or less independently of this deeper process. The other matter of widespread acceptance is that, whatever the immediate driving mechanism, the creation of new, hot oceanic lithosphere at constructive plate boundaries and the subduction and recycling of old, cold oceanic lithospere at destructive plate boundaries is the principal means by which the heat generated by radioactive decay in the mantle escapes to the surface. Probably about twice as much heat gets out this way as trickles out through the lithosphere by conduction, and each of these considerably outweighs the outward heat transfer associated with volcanic eruptions. One of the great unsolved mysteries of modern geology is why Earth behaves in this way but Venus, which is almost the same size, mass and density as the Earth, seems to lack plate tectonics.

There are many forces which could be causing plates to move. As already noted, simple conveyor-belt type drag by flow in the underlying asthenosphere is not considered likely. Nor does it seem that plates are pushed apart by forcible injection of new material along constructive plate boundaries. It is probably closer to the truth to regard the upwelling here as a *consequence* rather than a *cause* of plate divergence. Perhaps the most likely driving mechanism is that the old, cold edge of a subducting slab, sinking because it is negatively buoyant, drags the rest of the plate with it.

We have mentioned volcanoes several times so far. In the next chapter, we will consider them more fully.

5

VOLCANOES

We saw in the previous chapter that the world's most obvious volcanoes are sited in Andean mountain chains or island arcs above subduction zones. The global distribution of volcanoes above sea level (Figure 5.1) correlates well with destructive plate boundaries (Figure 4.9). There is rather more volcanic activity, though of a different kind and harder to detect because it is usually below several kilometres of water, occurring continually along constructive plate boundaries. In contrast, conservative plate boundaries rarely show associated volcanism.

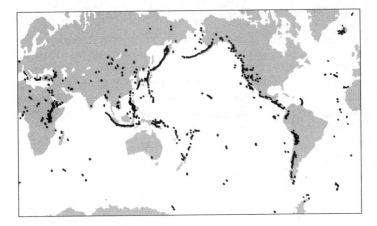

Figure 5.1 Global map of volcanoes above sea level, showing all volcanoes known to have erupted in the past 10 000 years.

In this chapter we will consider the types of volcanic activity that happen at plate boundary settings, and also look at some volcanoes that occur well away from the edges of plates.

Mention the word 'volcano', and most people probably think of something looking like the volcano in Figure 5.2; a fairly steep conical mountain like Mount Fuji in Japan. This sort of volcano, known as a stratocone, is typical of settings above subduction zones, though they are not all so symmetric.

Figure 5.2 A stratocone volcano, Acamarachi in northern Chile. The date of its most recent eruption is not known, but is believed to be within the past 10 000 years.

Magma origin and eruption

In this situation, the material that is being melted at depth belongs mostly to the crust of the descending slab. With the exception of any sediments that have accumulated on top of it, this will be basaltic in composition, implying a silica content of about 49% (Table 2.2). Most of the melt derived from this is richer in silica (about 55–60%) as a result of partial melting, a process we met in the previous chapter. Melts of the intermediate silica content produced by partial melting of basalt are called andesites (named after the Andes mountains, where they are common).

Magma is usually slightly less dense than solid rock, so once formed, it tends to rise to the surface. At first it percolates along the inter-faces between crystals, but eventually collects into larger blobs that can force their way upwards. Some magma bodies cool down suffi-ciently to solidify before reaching the surface, forming intrusions which we will look at in the next chapter. For now, we are concerned only with what happens if the magma reaches the surface. When it does so, one of two things happens. It may ooze out and begin to flow downhill. This is described as a 'lava flow', **lava** being the term used to describe magma once it has reached the surface. Alternatively it may explode to produce a **pyroclastic** (fire-broken) rock.

Magmatic gases and explosive eruptions

The idea of lava exploding may seem surprising. It is, however, a very common occurrence (Figure 5.3). What makes lava explode is the sudden escape of gases that have been dissolved within the magma. If these gases escape rapidly (like a can of beer opened after it has been shaken) bubbles form that expand the lava into a froth and may

Figure 5.3 An eruption from the south-east crater on Mt Etna, Sicily. Gases are escaping from lava within the vent with force sufficient to drive this mild pyroclas-tic eruption. Fine fragments are thrown to a height of about 50 m above the rim of the crater, and then blown downwind.

break it into fragments. Lumps of solidified bubbly rock are referred to as pumice. The small fragments are referred to as volcanic **ash**, which correctly describes their typical fine-grained and greyish colour but is potentially misleading because, unlike the ashes of a fire, volcanic ash is not a product of combustion.

Magma rising above subduction zones typically contains a few per cent of gas dissolved within it. Typical gas composition is about 56% H_2O (water vapour), 28% CO_2 (carbon dioxide), 14% SO_2 (sulfur dioxide), and 1% or less of each of H_2S (hydrogen sulfide) HCl (hydrogen chloride), H_2 (hydrogen) and CO (carbon monoxide). Above a subduction zone, most of the dissolved water vapour can be explained as sea water that had been transported down with the descending slab (in fissures in the oceanic crust or within wet sediments), and some of the carbon dioxide probably comes from breakdown of calcium carbonate (which is common in certain kinds of sedimentary rock). However, magmas in other settings also contain dissolved gas (though usually less) and it seems that some of the gas is escaping from within the mantle. This represents the tail-end of the degassing process that is thought to have been the source of most of the atmosphere (Chapter 2).

Whether a volcanic eruption is explosive does not depend on whether the magma is shaken like beer in a can, but on how quickly the magma rises and how easily the gases can escape. The great pressure at depth keeps gases dissolved in the magma when it is deep down, and usually bubbles begin to form only within a couple of kilometres of the surface. Events during the final ascent from this depth are crucial. If the magma rises slowly, then there will usually be time for it to degas quietly, and on reaching the surface it will form a lava flow. If it rises quickly, or if the walls of the conduit are impermeable to gas, the gas will remain trapped within the magma and it will erupt explosively.

In many volcanoes, gas escapes quietly over periods of tens or hundred of years without any associated eruption. The fissures where the gas reaches the surface are known as fumaroles, and may be at temperatures of up to several hundred degrees centigrade (Figure 5.4).

Eruption conditions in andesite stratocone volcanoes tend to fluctuate between explosive and quiet. The resulting alternation of lava flows and more widely dispersed ash layers is chiefly responsible for the relatively steep conical shape of this kind of volcano (Figure 5.5)

Figure 5.4 Volcanologists studying fumaroles on the summit rim of Vulcano, a volcanic island to the north of Sicily that gave its name to volcanoes in general. On reaching the atmosphere the volcanic water vapour forms a steam plume. The volcanologists are wearing gas masks to filter out the acid gases, in case the wind shifts.

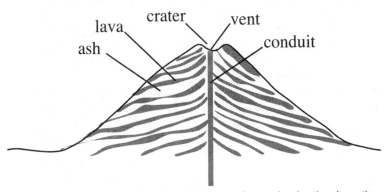

Figure 5.5 A cross-section through a stratocone volcano, showing the alternation of ash and lava responsible for its shape. There is only a single vent shown here, but in many cases the conduit branches at depth, feeding subsidiary (parasitic) vents on the flanks of the volcano.

Volcanic hazards

Lava flows of andesite composition are fairly viscous. They tend to have very blocky surfaces, and move more like a train of rubble than a stream of liquid. As a result, they are usually shorter than about 10 km in length. Although they may damage property, they rarely kill people because they mostly advance at only a few metres per hour.

When andesitic volcanoes erupt explosively, the effects are much more widespread. The example shown in Figure 5.3 is at the mild end of the scale. One of the biggest eruptions of the twentieth century occurred in June 1991 at Mt Pinatubo in the Philippines. This volcano had not erupted for at least 500 years. As is often found, when it awoke it erupted much more dramatically than a similar volcano in the habit of erupting every few years, and a column of ash reached a height of about 60 km over the volcano. This type of eruption is described as plinian, after Pliny the Younger whose description of a similar eruption of Vesuvius in 79 AD, which led to the destruction of the towns of Pompeii and Herculaneum (and the death of Pliny the Elder, his uncle), is the first stage-by-stage account we have of a volcanic eruption. A plinian eruption column of an intermediate size is shown in Figure 5.6.

Figure 5.6 A plinian eruption column rising about 20 km above Lascar volcano, in the Andes. (Courtesy of Stuart Hall, British Antarctic Survey.)

In a plinian eruption, the ash is projected upwards by the force of the explosive escape of gas near the vent. However, columns of ash would not attain the great heights that they reach without the aid of buoyancy. What happens is that air is drawn into the column and heated by the hot ash particles and by mixing with the hot volcanic gases. The heated air expands, and, despite the weight of the ash particles, the column becomes buoyant and rises until it reaches a height where its density matches that of the surrounding air. The eruption cloud is then blown downwind, and the ash begins to settle out to form an 'airfall' ash desposit. The larger particles fall faster, and the airfall deposits tend to be thicker and made of larger particles closer to the volcano (Figure 5.7). The largest bits, ranging up to metres in size,

Figure 5.7 The volcanic ash here seen in cross-section on the island of Vulcano represents several eruptive pulses, in each case the larger particles falling first so that each layer gets finer upwards.

are known as volcanic bombs. By mapping out the thickness of ancient airfall deposits it is possible to determine the direction of the wind at the time of the eruption. Where it accumulates thickly, airfall ash can cause roofs to collapse, and the fine particles can cause choking, but it is rarely a major cause of death.

Sometimes, an eruption column becomes unstable and all or part of it collapses onto the side of the volcano, and then sweeps downslope across the ground surface as a cloud of searing hot ash and gas moving at over 100 km per hour. This sort of pyroclastic flow is known as a nuée ardente (French for 'glowing cloud') and can also be triggered by the collapse of a steep dome of extruded lava. A nuée ardente is justifiably the most feared consequence of a plinian eruption. One reason for this is the destruction of the town of St Pierre, the capital of the Caribbean island of Martinique, and the death of all but two of its 29 000 inhabitants. One survivor was a prisoner locked in the town jail, in a cell without a window and partly below ground level. He was dug out badly burned two days later, and pardoned for his crimes. Fears of a similar eruption on the nearby island of Montserrat led to the evacuation of its most vulnerable areas in 1995 and 1996.

Unlike airfall deposits, which blanket the topography irrespective of slope, pyroclastic flows tend to be confined to valleys; a characterisic that enables the two types of deposit to be distinguished. However, not all airfall material remains in place. In particular it is easily washed away when it rains. In tropical areas the rain is liable to be torrential, so the water courses fill with a slurry of ash and water that, being denser than water, moves with a great force. These volcanic mudflows (often called lahars; an Indonesian word) caused most of the damage resulting from the 1991 eruption of Mt Pinatubo. Lahars can also be initiated if a volcanic eruption melts a glacier or snowcap. This is what happened during an eruption of Ruiz, Colombia, in 1985 when approximately 23 000 people, including most of the population of the town of Armero, lost their lives. Readers may remember harrowing television pictures of a young girl, trapped by solid debris in the mudflow, and with only her head showing, who succumbed after several days' attempts to dig her out.

The principal hazards deriving from explosive eruptions of andesitic volcanoes are summarised in Figure 5.8. In addition, we may add the risk of volcanic collapse, a process that first became widely-recognised after the May 1980 eruption of Mt St Helens, in Washington State, USA. Here there was two months' warning of a major eruption in the form of small but progressively shallower earthquakes, minor ash eruptions, and a slight bulging of the northern flank of the volcano. However, rather than culminating in a plinian eruption from the summit crater, the northern flank gave way, forming a giant debris avalanche. As this collapse began, the confining pressure on the magma within the volcano was released and it degased explosively, directing most of its force sideways. This directed blast was superson-

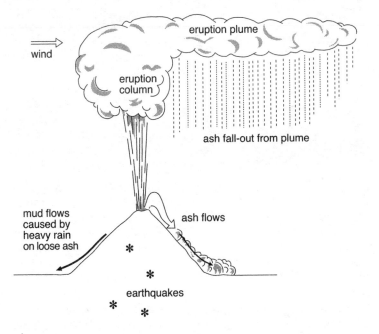

Figure 5.8 Some of the hazards associated with explosive eruptions. The most dangerous ash flows are nuées ardentes.

ic, and 60 square kilometres of what had been forest was devastated (Figures 5.9 and 5.10).

The sideways nature of this directed blast caught even the professionals by surprise. A volcanologist working for the US Geological Survey, Dave Johnson, was stationed in what should have been a safe location, on a high ridge several kilometres north of Mt St Helens, and was assigned the duty of recording the anticipated eruption. Tragically, his camp site caught the full force of the directed blast, and his body was never found. His young field assistant, Harry Glicken, had been sent back to headquarters, and survived to become a well-known volcanologist in his own right. Sadly he too died, eleven years later, when he was caught by a nuée ardente generated by a collapsing lava dome on the Japanese volcano Unzen.

The Mt St Helens collapse was triggered by the injection of magma

Figure 5.9 Mt St Helens, three years after its 1981 eruption. The whole summit was undermined and destroyed by the collapse and directed blast. The steam within the collapse amphitheatre comes from a slow-growing lava dome that may eventually restore the volcano to its previous symmetrical cone shape.

Figure 5.10 These mature Douglas firs were blown sideways, stripped of their branches and sharpened to points by the directed blast from Mt St Helens, May 1981.

into the volcano, but it is possible that other volcanoes collapse simply under their own weight. When this happens debris avalanches can travel for tens of kilometres, to judge from mapping of ancient deposits. No major volcanic debris avalanche has happened in historic times.

Other causes of death associated with volcanoes include the release of suffocating gases, which may flow downhill, and tsunamis generated not by earthquakes but by underwater volcanic explosions, or by a volcanic landslide entering the sea or a lake, Table 5.1 lists the fatalities from some notable volcanic events.

Volcano	Year	Fatalities	Main cause(s) of death
Vesuvius, Italy	79AD	>3500	Pyroclastic flows
Kelut, Indonesia	1586	10 000?	unknown
Asama, Japan	1598	800	Religious pilgrims killed at summit
Vesuvius, Italy	1631	>4000	Pyroclastic flows
Merapi, Indonesia	1672	3000?	Pyroclastic flows
Laki, Iceland	1783	9350	Starvation
Asama, Japan	1783	1500	Pyroclastic flow, lahars
Unzen, Japan	1792	14 300	Tsunami
Tambora, Indonesia	1815	92 000	Starvation
Krakatau, Indonesia	1883	36 400	Tsunami
Mt Pelée, Martinique	1902	29 000	Pyroclastic flows
Taal, Philippines	1911	>1335	Pyroclastic flows
Merapi, Indonesia	1930	1369	Pyroclastic flows
Ruapehu, New Zealand	1953	151	Lahar
Iliwerung, Indonesia	1979	539?	Tsunami
Mt St Helens, USA	1980	57	Directed blast, lahars
Mayon, Philippines	1981	>200	Lahar
El Chichon, Mexico	1982	1900	Pyroclastic flows
Ruiz, Colombia	1985	23 000	Lahar
Lake Nyos, Cameroon	1986	>1700	Asphyxiation by gases
Pinatubo, Philippines	1991	800	Airfall, lahars, disease

Table 5.1 Some notable volcanic events.

Not all volcanoes above destructive plate boundaries are andesitic stratocones. In some places basaltic melts approach the surface, either because of more complete melting of the subducting slab of oceanic crust, or because of partial melting within the wedge-shaped volume of mantle between the two plates. This basalt may flow out as quiet lava flows, its lower viscosity allowing it to spread more thinly than andesite over many square kilometres, or if the eruption is driven by the force of escaping gases it may result in a 'cinder cone', formed of lumps of slightly frothy basalt (Figure 5.11).

Figure 5.11 A basaltic cinder cone, in the Andes of northern Chile.

On the other hand, very large volumes of more silica-rich magmas are sometimes generated above subduction zones. This is particularly likely where large proportions of sediment have been dragged down by the subducting slab, or when there is a lot of melting of the lower crust above the subduction zone. These magmas typically have around 70% silica, and are described as acidic or granitic. These usually crystallize at depth, forming the well-known rock-type called **granite**, which we shall examine in the next chapter, but sometimes they do approach the surface.

It is rare for granite magmas to ooze out quietly. This is because granitic lava is even more viscous than andesitic lava, making it difficult for gases to escape quietly. Sometimes it does happen though, and a very thick flow or a steep-sided lava dome is formed (Figure 5.12)

Figure 5.12 An image of a 40 km wide area in the Andes, recorded by satellite. On the left is a 24 cubic km acidic flow, whose front is 300 m high. Its thickness and the large-scale wrinkles on its surface indicate that this was an extremely slow-moving viscous flow. On the right is an acidic dome, only one-tenth the volume, which appears to have oozed out radially from a vent hidden below its centre.

Ignimbrite eruptions

What happens more often is that a depth is reached, maybe a few hundred metres below the surface, at which gas pressure fractures the roof over the magma body, thereby releasing pressure and encouraging whatever gas remains in solution to come out all at once. This drives the most energetic variety of explosive eruption known. Such eruptions are known from study of ancient deposits rather than observation because (fortunately for us) they are infrequent and no large example has occurred in recent times. Probably a very high plinian eruption column develops, to judge from the airfall deposits that we see preserved. The most characteristic deposit, though, is probably a result of column collapse. This takes the form of a very extensive sheet of acidic ash, in which the particles are often welded together because they were still very hot when emplaced. This sort of deposit is known as an ignimbrite. There are good recent examples in the south-west USA (Figure 5.13), New Zealand and the Andes, and 400-million-year-old examples in Snowdonia (Wales) and the Lake District of England.

Figure 5.13 These 50 m high cliffs expose a section through an ignimbrite that was erupted from the Valles caldera, New Mexico, USA.

Large ignimbrites are the result of eruption of up to around a hundred cubic kilometres of magma. Such an eruption tends to leave a large hole in the ground, rather than an obvious volcanic mountain. Usually what had been the roof over the magma body collapses to form the floor of a large crater, termed a caldera (Figure 5.14).

Figure 5.14 Cross-section to show caldera formation. Between these two stages, a large-volume ignimbrite-forming eruption occurs.

I have described volcanism near destructive plate boundaries in broad outline. Unfortunately, we cannot explore the many details of eruptive processes in a book of this length. We will look now at volcanism at constructive plate boundaries.

Volcanoes associated with continental rifting

First, what happens when a continental plate is in the process of splitting to initiate a new constructive plate boundary? Figure 4.6 showed that this happens above a linear zone of upwelling in the asthenosphere. The faults that form and allow the crust to stretch are actually rather good pathways along which magma can rise upwards, and indeed continental rifting seems usually to be preceded by an approximately ten-million-year period of eruption of mainly basaltic lavas, produced by partial melting of the upwelling mantle asthenosphere. This is going on today in the African Rift, a large fault-bounded structure that runs from Eritrea and Ethiopa southwards through Kenya and Tanzania to Mozambique. Africa appears to be splitting apart by east–west extension across this rift, although there is no guarantee that this will progress far enough to generate a new ocean.

This extension provides the upward pathways for the magma feeding the large African volcanoes like Kilimanjaro. Asthenospheric upwelling is at its most intense near the northern end of the African Rift, corresponding to a **hot spot** over a plume rising from deep within the mantle, below the point where the rift joins the Red Sea and the Gulf of Aden. Here the eruptions have taken the form of vast sheets of basaltic lava flows, known as continental flood basalts.

There are many ancient examples of flood basalts that appear to represent hot spots prior to continental rifting. One of the largest and most famous is the 65-million-year-old Deccan Traps, in India. These are a kilometre thick and still cover an area of half a million square kilometres, and are signs of a hot spot that helped India to rift away from Africa and Madagascar. (The hot spot is still active, being marked by the site of the volcanically active island of Reunion). Other continental flood basalt provinces reflect the rifting of the Atlantic ocean, and can be found split into unequal parts on either side. These include: the 125-million-year-old Parana basalts of southern Brazil and their less-extensive counterpart the Etendeka traps in Nambia; and the smaller but more famous 60-million-year-old Brito-Arctic province, most of which is in east Greenland, but which includes the thick lava flows of the Giant's Causeway, in Northern Ireland, and the Inner Hebrides islands, Scotland (Figures 5.15 and 5.16).

The 15–17-million-year-old Columbia River Basalts of Washington, Oregon and Idaho constitute a continental flood basalt province above a

Figure 5.15 The island of Staffa, formed of a 60 m thick basalt flow, the middle part of which (from present sea level to the roof of the cave) fractured into columns as a result of contraction during comparatively slow and orderly cooling. The upper part is more rubbly. The cave is Fingal's Cave, made famous by Mendelssohn in his overture The Hebrides.

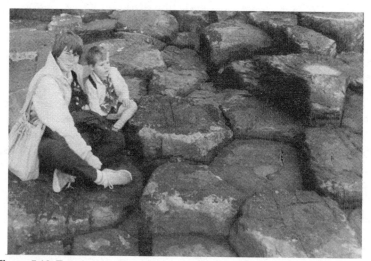

Figure 5.16 Top view of columnar joints formed by slow cooling within the plateau basalt flow on Staffa. Most columns are hexagonal, though five-sided examples can be seen.

hot spot that has not been followed by a major rifting event. One of these flows travelled 300 km along the Columbia River Gorge before spreading out laterally over 100 km. It has been suggested that the whole volume (over 700 cubic km) may have been erupted in just one week!

Volcanism at ocean ridges

Once true ocean crust has begun to form, by sea-floor spreading, the character of the volcanism changes. The basaltic lava erupted at ocean ridges to form the top layer of the oceanic crust takes on a form completely unlike that of lava on land. When lava is extruded under water, its surface chills very rapidly, forming a flexible rind around the still-molten interior. The lava resembles a pile of water-filled plastic bin-liners, a morphology known as '**pillow lava**' (Figure 5.17).

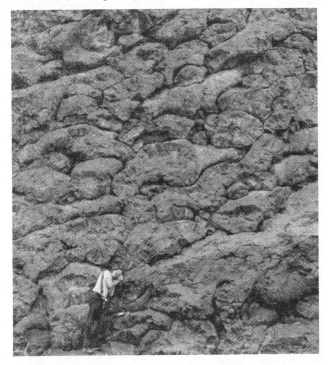

Figure 5.17 Pillow lava, erupted on the ocean floor 95 million years ago at a constructive pate margin. This is now exposed on land as part of an ophiolite in the Oman mountains, Arabia, being a rare example of a slice of oceanic lithosphere that instead of being subducted was thrust on top of continental crust during a collision.

Where lava is erupted fast enough at a constructive plate margin, it may form sheet flows. Alternatively it may fragment to form a glassy debris known as hyaloclastite. However, explosive eruptions are virtually unknown because the pressure exerted by the overlying depth of water inhibits the escape of gases.

One special characteristic of volcanism at constructive plate margins is the widespread occurrence of vents where superheated water emerges. Water percolates down through the crust over a wide area, is warmed within and below the lava layer by heat from the underlying magma and is expelled, by convection, through vents along the axis of the ridge. On the way it dissolves some elements from the rock, and certain other elements already dissolved in seawater come out of solution by reacting with the rock. The hot water emerges as an acidic solution, but as soon as this mixes with seawater various of its dissolved constituents become insoluble, and form a smoke-like cloud of metal sulfide particles. The most famous examples are vents at about 350 °C where the particles are black, hence their name 'black smokers'. These hot vents are surrounded by organisms like clams, crabs and worms that ultimately feed not off plants that derive their energy from sunlight, but off bacteria deriving their energy by oxidising the sulfide particles. It has been suggested that life on Earth (and elsewhere) could have originated in such a setting, and only later spread to sunlit regions where plants learned to photosynthesise.

We noted in the previous chapter that the partial melting at a constructive plate margin is caused by the decompression associated with the upwelling of asthenosphere at a rate necessary to replace the lithosphere drawn away by subduction. However, when a constructive plate boundary is situated above a plume coming up from deep within the mantle, the volume of melt produced is much greater than usual. As a result the oceanic crust may be two or three times thicker than normal, and its surface will be above sea level. The only really clear example of this today is Iceland.

Hawaiian volcanism

We turn now to volcanoes that are unrelated to plate boundaries, and consider the example of Hawaii. A plume from deep within the mantle is involved again, but this time the plume hits the base of the lithosphere more than 1000 km from the edge of the Pacific plate. This plate is moving north-westwards over the plume, and as a result there is a chain of increasingly older, extinct volcanoes stretching north-westwards from the Big Island of Hawaii where the current activity is centred. With time, the weight of these older volcanoes causes them to subside, and they are also worn down by erosion, so only the youngest few remain above sea level, the rest of the chain surviving only as undersea peaks known as seamounts.

During the most active part of its lifetime, a Hawaiian volcano erupts nothing but basalt. Basalt is the least viscous type of lava. There are two important consequences of this for a volcano above sea level. First, any gas within the lava can escape relatively easily, so eruptions tend not to be highly explosive. Second, the lava flows, being so runny, are able to spread out thinly. In consequence, Hawaiian volcanoes are characterised by gentle slopes of 2–3 degrees. Such an edifice is known as a **shield volcano**, because in profile it resembles a shield laid on the ground (Figure 5.18). When lava flows reach the coast their fronts get chilled by the seawater, so they tend to spread sideways along the shore. This steepens the shoreline to a point where it collapses, and much of that portion of the cone below sea level is made of collapsed lava rubble rather than intact lava flows.

50 km

Figure 5.18 The profile of a Hawaiian-type shield volcano. Mauna Loa volcano on the Big Island counts as the highest volcano on Earth, because it rises over 9 km from its base on the floor of the Pacific Ocean.

Basalt lavas like those erupted on Hawaii take essentially two forms. When moving relatively fast, because of a fast eruption rate or a steep slope, the surface breaks into clinkery blocks. This type is known by the Hawaiian term 'a'a' (pronounced 'ah-ah', and said by some to represent the expressions of pain made by someone trying to walk barefoot over such a jagged surface). When moving more slowly, a smooth, chilled but pliable surface forms over the molten interior of the flow. This does not fragment as in an a'a; instead, drag by the flowing liquid lava beneath tends to crumple the surface over upon itself to form a texture reminiscent of coiled rope. This ropey lava is often described by another Hawaiian word, pahoehoe. Both kinds of lava are visible in Figure 5.19.

Figure 5.19 Two kinds of basalt lava in Hawaii: clinkery a'a (left) and ropey pahoehoe (right).

Lava hazards

Basaltic lava flows, even though they are the least viscous and therefore fastest flowing of common lava types, usually advance at less than five metres per second. They rarely kill people, because there is usually plenty of time to get out of the way. A tragic exception occurred when the wall of a lava lake on the volcano Nyiragongo in Zaire became breached and the lava escaped at an average speed of 30 km per hour, taking the lives of at least fifty people. Normally, though, it is property and livelihoods rather than lives that are threatened by basaltic volcanism. Lava flows are virtually unstoppable, and easily overcome barriers erected to try to pond or divert them. One of the more notable successes in lava flow diversion was in 1992 when the front of a major lava flow on Etna, Sicily, had reached within a few hundred metres of the town of Zafferana. The side of the flow was deliberately breached by explosives part way along its length, which caused the flow to spread out sideways high up on the mountain, and halted its downstream advance.

Predicting eruptions

About sixty volcanoes erupt each year. A tenth of these eruptions are over within a day and some go on for years, but the average duration is about seven weeks. How can we tell when an eruption is due, and how can we predict what its effects will be?

The best way to anticipate an eruption's effects is to study previous eruptions. This is easy in the case of volcanoes that erupt every few years, because there are likely to be eye-witness reports, video recordings, and so on. However, it is the volcanoes that erupt least frequently whose eruptions are usually the most dangerous, because a greater volume of ash and/or lava tends to come out in one go. In such cases, there may be no historical records, and the volcanologist must depend largely on the traditional skills of the field geologist – specifically, interpretation of past events based on detailed mapping of ancient deposits. In addition, computer programs exist that can use a digital model of the terrain to predict the paths of ash, lava and mud flows. Hazard zone maps are available for many volcanoes, based on these approaches.

There are many precursor signs of an impending eruption. If recognised, these can give days' or even months' warning. In this respect volcanoes are easier to predict than tectonic earthquakes in which, as we have seen, the first shock is usually the biggest. Probably the single most useful eruption precursors are special kinds of small earth tremors, readily distinguished from tectonic earthquakes, whose depth typically decreases from several kilometres to a few hundred metres during the build-up to an eruption. These are associated with injection of magma into the volcano. In order to pinpoint them it is necessary to have an array of at least three seismometers on or close to the volcano.

Seismometers are expensive instruments, and few volcanoes have permanent arrays. At the time of the first historic eruption of Pinatubo, Philippines, in 1991 the volcano had been little studied and was not under surveillance until a small ash eruption near the summit in March showed that it was waking up. An array of portable seismometers was then rushed into place, enabling the locations of sub-volcanic tremors to be tracked as they approached the surface. This gave enough notice for the most vunerable areas to be evacuated before the catastrophic eruption in June. Most of the ensuing deaths were attributable to the failure to sweep airfall ash off flat roofs (which collapsed under the weight) and subsequent mudflows, and not due to lack of eruption warning.

When magma is forced into a volcano, the volcano swells in size. In extreme cases its contours may change by over a metre. This deformation is another useful precursor sign, and its pattern may serve to indicate whether the eruption is likely to occur at the summit or from a vent or fissure on one side. It can be studied by ground-based surveying or by using radar data from satellites.

Other techniques include measuring tiny changes in gravity that reflect injection or withdrawal of magma from a volcano, measuring changes in the rate of gas emission or of gas composition, and monitoring changes in surface temperature using ground-based instruments or satellite-borne infra-red detectors.

Widespread effects of eruptions

Volcanic eruptions don't just affect people living locally. Several intercontinental airliners have run into trouble by unknowingly flying into volcanic ash clouds, whereupon their engines have stalled. Luckily, on every occasion so far the pilots have been able to restart the engines, though sometimes only after a terrifying gliding descent most of the way to the ground. Eruption cloud warning systems, based on images from weather satellites, and specially designed aircraft-mounted radar are now coming into place to protect the most vunerable air traffic routes, notably the trans-polar route between Europe and Japan that passes over Alaska and the Kuriles, and routes to Australia via Indonesian airspace.

The biggest volcanic eruptions can have continental or even global effects. The 1991 eruption of Pinatubo injected so much fine ash and sulfur dioxide into the stratosphere that global temperatures are estimated to have been reduced by about half a degree centigrade throughout the following year. This is because sunlight that would normally have warmed the surface was blocked in the upper atmosphere. The largest eruption in modern history was of Tambora, Indonesia, in 1815. About 50 cubic km of ash was erupted, which was responsible for glorious red sunsets round the globe and a much greater drop in temperature than after the Pinatubo eruption. In the northern hemisphere, 1816 has been described as 'the year without a summer'. There were widespread crop failures and probably more deaths than from any other historic eruption.

When we consider the likelihood of a similar eruption in the near future, not to mention an even larger ignimbrite-forming event, we have to accept that volcanoes could play a big role in short-term climate change. The possibility of world-wide famine cannot be discounted.

Volcanic activity also has more benign effects. Magma at shallow depths can heat the ground water, and cause it to gush episodically through a vent in the form of geyser, like the archetypal Geysir in Iceland or Old Faithful in Yellowstone Park, Wyoming. Heated water may also seep out gently to form hot mud pools. Bathing in these is said to be therapeutic, though personally I have found the clinging smell of rotten eggs (caused by hydrogen sulfide) to be an annoyance that seems impossible to get rid of, no matter how long I shower for afterwards!

Hot springs are as much a manifestation of magma intruded at depth as of magma erupting at the surface, and we will look at these intrusions in the next chapter.

6

IGNEOUS INTRUSIONS

Rocks formed by cooling of a magma are called igneous rocks, from the Latin word for fire. We were dealing with igneous rocks of the volcanic kind throughout the previous chapter. However, not all magma reaches the surface. Often it solidifies at depth, to form an **intrusive** igneous rock.

Volcanic rocks cool so quickly that usually the only crystals that grow within them are too small for the unaided eye to see. However, an intrusive rock cools more slowly, so there is time for the crystals of which the rock is formed to grow to lengths of millimetres or centimetres before the rock is completely solidified. Crystals this big are easy to see when the body is eventually exposed on the surface as a result of erosion. It is time now to consider the crystalline nature of igneous rocks, before looking at the sorts of intrusive bodies that they can form.

Minerals

Because the crust and mantle contain so much silicon and oxygen (Table 2.2), all the common sorts of crystals that grow in magmas from these sources contain these elements. The crystalline structure of rock-forming igneous minerals is based on two- or three-dimensional arrays of silicon and oxygen held together by molecular bonds. These minerals, as well as the rocks that they form, are commonly referred to as silicates.

The most familiar of the silicate minerals is pure silica. This contains just silicon and oxygen, and has the formula SiO_2, which in crystalline form is the mineral quartz, except at extremely high pressures when the atoms are squeezed into more tightly-packed crystalline structures. All other silicate minerals contain various metallic elements in addition to silicon and oxygen. Quartz is a robust, hard-wearing mineral whose significance you will realise in Chapter 7. It is able to grow in magmas whose SiO_2 content exceeds about 52%. Quartz is abundant in rocks of granitic composition and is usually present, though scarce, in rocks of andesitic composition. The potentially confusing use of silica to mean two quite different things is standard in geology. The silica (or SiO_2) content quoted for a rock is based on an overall chemical analysis of the elements making up the rock. This should not be taken to imply that all or any of the silicon and oxygen occurs in the form of quartz or any other pure silica mineral.

On the other hand, when SiO_2 is given as the chemical formula of a mineral, it means that this mineral contains Si and O only, in the ratio 1:2. This specific ratio is dictated by number of oxygen atoms each silicon atom is able to form a bond with. The three-dimensional structure of such bonds controls the properties of the crystal.

In magmas whose composition is basic (45–52% silica) or **ultrabasic** (<45% silica), metallic elements are so abundant that no quartz crystals can form, and the only silicate minerals to form contain a relatively high proportion of metals. The most extreme example is olivine, formula Mg_2SiO_4, in which there are twice as many metal atoms (in this case, magnesium) as silicon atoms. Actually, iron (Fe) atoms can fit in the same place as magnesium atoms, so olivine can have any composition ranging between Mg_2SiO_4 and Fe_2SiO_4, with Mg and Fe in any proportion. The permitted variation in olivine's formula is often expressed by quoting it as $(Fe,Mg)_2SiO_4$. This degree of variability in composition, which is achieved without changing the internal structure of the crystal, is what distinguishes a mineral from a chemical compound. Usually in igneous rocks olivine is of a magnesium-rich variety, say $Fe_{0.2}Mg_{1.8}SiO_4$.

Olivine most often occurs in association with pyroxene, another metal-rich silicate, with the formula $(Ca,Mg,Fe)_2Si_2O_6$. A rock composed essentially of just olivine and pyroxene is described as a peridotite, and these are the dominant minerals in mantle rocks.

In magmas of higher SiO_2 content, the proportions of magnesium and iron tend to be lower, whereas aluminium (Al), calcium (Ca), sodium (Na), and potassium (K) become more abundant. One result is the appearance of minerals known as feldspars. Plagioclase feldspar, the most abundant mineral in rocks of andesitic composition (52–66% SiO_2), has a formula ranging between $NaAlSi_3O_8$ and $CaAl_2Si_2O_8$. Potassium feldspar ($KAlSi_3O_8$) occurs chiefly in granite rocks (>66% SiO_2), in association with quartz and plagioclase feldspar.

The other important silicate minerals are mica and amphibole. These occur in granite and andesitic rocks and depend on the presence of water in the magma. They have complicated formulae: muscovite mica is $KAl_2(AlSi_3O_{10})(OH)_2$, biotite mica is $K(Mg,Fe)_3AlSi_3O_{10}(OH)_2$, and amphibole is $(Na,K)Ca_2(Mg,Fe,Al)_5(Al,Si)_8O_{22}(OH)_2$.

Each of these minerals, and varieties within each type, can be distinguished by features such as colour, hardness, **cleavage** (regular planes of weakness in the crystal resulting from its atomic structure), and optical properties. A few characteristics of the main minerals are summarised in Appendix 1. With practice it is possible to recognise many of these when you find a coarse-grained igneous rock outcropping in the field, without the aid of analytical instruments. Having identified each of the minerals making up a rock, it is then possible to give the rock itself a name, based on the criteria summarised in Appendix 2, and thence to deduce the conditions under which the rock is likely to have formed.

We will say no more about the mineral building blocks of igneous rocks here. The collection and study of minerals is an enthralling pastime in its own right, and there are many books on the subject.

Crystallisation of minerals in cooling magma

As a magma rises it gets cooler, because heat escapes into the surrounding rocks. Eventually the temperature and pressure will be appropriate for minerals to start crystallising. The nature of these minerals depends on the overall composition of the magma. The growth of the mineral crystals is not like the freezing of water, where the whole liquid solidifies at a specific temperature. Instead, each mineral species begins to crystallise at a different temperature. Generally speaking, the more metal-rich minerals crystallise at higher temperatures, and quartz and the sodium- and potassium-rich feldspars crystallise at lower temperatures. This process, which is really partial melting in reverse, is known as **fractional crystallisation.**

If the crystals are carried along with the remaining magma until everything has crystallised, then the overall composition of the rock is the same as that of the initial magma, even though the rock consists of several minerals each having a different composition. However, if the first crystals become separated from the magma (perhaps because they settle to the bottom, or because the magma gets squeezed out) the remaining magma will have a different composition, richer in silica, from what it started with. In this way andesitic magmas can evolve from originally basaltic magmas, and granitic magmas can evolve from andesitic magmas (Figure 6.1). When these new magmas crystallise, they will form rocks of different composition to that of the initial magma.

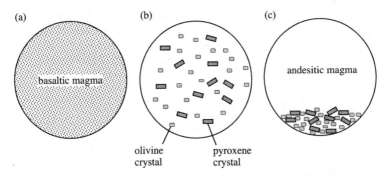

Figure 6.1 Schematic diagram, not to scale, showing how a body of basaltic magma (a) can begin to grow crystals of olivine and pyroxene (b). If these crystals settle out, as in (c), they form a rock of ultrabasic composition and the remaining magma (which may crystallise later) is andesitic in composition.

Near the surface, a basaltic magma begins to crystallise at about 1200 °C and is entirely solidified by about 1080 °C. The corresponding temperatures for a granitic magma are 1100 °C and 960 °C, and those for an andesitic magma lie somewhere in between. At depth, the pressure acts to increase these temperatures by about 20 °C for every 10 km depth. However, this applies only for water-free magmas. Above destructive plate boundaries magmas are usually rich in water. The effect of this is to reduce the temperature at which crystallisation begins, so that in a 'wet' granitic magma at 10–30 km depth crystals do not begin to form until the temperature has dropped to about 720 °C. On approaching the surface, the water vapour is liable to escape, as we saw in the previous chapter. If the magma had already cooled to close to its 'wet' crystallisation temperature, then this loss of water will cause the magma to solidify, because it will be well below its 'dry' crystallisation temperature.

Granite

This is one reason why granitic magmas in particular are more widely known for the large intrusions they form than for volcanic products. Another reason is that they are about a thousand times more viscous than basalts, and so are less easily extruded onto the surface. It is not uncommon to find coarse-grained solidified intrusions of granitic composition that are tens of kilometres across. These constitute the well-known rock type granite. In fact, the name granite is so well known that monumental stone masons and the like tend to use it for any crystalline rock, whether or not it has the mineralogical composition that a geologist would call granite. Poets use the term even more loosely, to refer to any rock that they consider to be hard and enduring, although (as we will see in Chapter 8) granite is far from immune to the effects of wind and rain.

How then does a granitic magma, originating at, say, 30 km depth near the base of the crust, rise upwards? As magma forms it seeps along grain boundaries, but cannot flow freely enough to escape until about 5% of the source rock has melted, which may take ten thousand years or more. If there is a nearby plane of weakness such as a fault or fracture the magma may escape up it in a matter of a thousand years or so and spread out higher in the crust to form a large granite mass. However, if there is no easy pathway available magma will remain trapped at depth until it has grown into a large enough body to force its way up to shallower level by buoyancy alone. This is thought usually to require melting of about a third of the source rock. At first, the surrounding rocks will be very hot, and soft enough to allow the granite magma to push them aside, rising at a rate of a few centimetres or metres per year. As the granite reaches progressively shallower depths the surrounding rocks will be colder and therefore more brittle so that eventually the granite can rise no further.

There are two other processes that may aid magma ascent. First, if the overlying rocks have a melting point similar to or lower than that of the magma, they may become assimilated into the magma. Essentially, the magma melts its way through, continuing to rise until it has run out of heat. This, incidentally, is another way in which magma composition can evolve. Alternatively, the granite may pluck off blocks of rocks from above. These sink down through the granite, allowing the granite to pass upwards. This process is called 'stoping', and is shown in Figures 6.2 and 6.3. The stoped blocks, which can be of any size, may either sink to the bottom of the magma unscathed, or may be melted by the heat of the magma and become incorporated within it.

Most of the granites with which geologists are familiar ceased to rise at depths of a few kilometres within the crust. They have become exposed at the surface, millions of years after their emplacement, when the original overburden or rock has been worn away. This is encouraged by the fact that, even when solid, granite is less dense

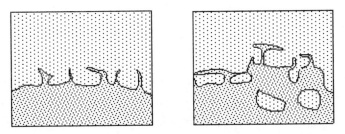

Figure 6.2 Sketch to illustrate stoping at successive intervals. A rising body of magma plucks blocks off the roof to make way for itself. The scale could be anything from one metre to several hundred metres across.

Figure 6.3 A pale granitic rock that became solidified in the act of stoping its way up through a darker rock type. The geological hammer gives the scale.

than most other rock types, and it tends to buoy up the terrain, increasing its vulnerability to the processes of erosion.

Granites can form wherever the composition of originally basaltic or andesitic magmas has been able to evolve so as to become sufficiently rich in silica, and so they can be found in any setting. However, they are particularly common above destructive plate margins (caused in particular by the melting of the lower crust above a subduction zone) and in the mountain belts at sites of continental collision. Here, the thickness of the continental crust is usually doubled (Figure 4.3), and because uranium, thorium and potassium are concentrated in the crust, the rate of local radiogenic heat production is doubled too. This encourages melting low down in the thickened crust even after relative motion between the two collided plates has ceased.

Large granite massifs such as that shown in Figure 6.4 are termed 'plutons', after Pluto, the Greek and Roman god of the underworld. Sometimes, geophysical evidence suggests that several plutons are joined together at depth, in which case the larger unit is referred to as a 'batholith'. The relationship between plutons and a batholith is shown in Figure 6.5.

Figure 6.4 Half Dome, a famous granite pluton exposed in Yosemite National Park, California. The sheer face, beloved of climbers, is a result of erosion being controlled by cooling joints.

IGNEOUS INTRUSIONS

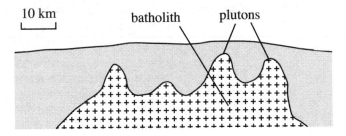

Figure 6.5 Cross-section through the crust, showing several granite plutons as upward projections from a more extensive batholith at depth.

Not all plutons are granitic in composition. For example, the 60-million-year-old continental flood basalts that we met in the previous chapter (reflecting the rifting of the north Atlantic ocean between Britain and Greenland) are associated with plutons of basaltic composition. Being course grained, we describe them as gabbro rather than basalt (see Appendix 2).

Basaltic intrusions in the oceanic crust

Gabbro is also crystallising to form the lower oceanic crust at constructive plate boundaries. This gabbro layer can be recognised seismically in the oceans and can be studied on-land in ophiolites, which you may remember from Chapter 4 are slices of oceanic crust and upper mantle that have been thrust over the edge of a continent during a collision. It was once thought that the magma chambers within which these oceanic gabbros crystallise are ten or more kilometres wide. However, studies in the ocean attempting to map seismic reflections from the roof of these chambers, or to locate the molten volumes where S-waves cannot penetrate, have demonstrated that these magma chambers are narrow, and possibly ephemeral, features. Much of the gabbro appears to arrive in place as a crystal-rich mush.

Between the gabbro layer and the basaltic pillow lavas erupted on the sea floor, there is a layer that is neither plutonic nor volcanic. This marks the depth at which pulses of magma are injected upwards along a vertical fissure to feed the lavas. When magma freezes in such a fissure it forms a vertical sheet of intrusive igneous rock called a **dyke**. In sea-floor spreading, successive dykes intrude each other, and so a layer is formed consisting of nothing but dykes, which is sometimes called a sheeted dyke complex. The relationship between these layers in the oceanic crust are shown in Figure 6.6.

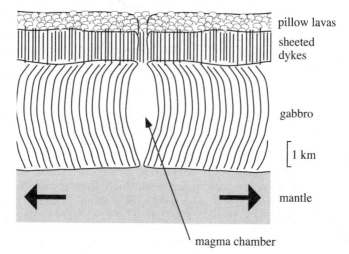

pillow lavas

sheeted dykes

gabbro

1 km

mantle

magma chamber

Figure 6.6 The constitution of the oceanic crust, showing a cross-section through a spreading axis. Crust on either side is moving apart. Basaltic magma cools slowly at depth to form the coarse-grained basaltic rock type known as gabbro. The curved lines on this figure represents successive walls of the magma chamber which can maintain a constant size if the rate of the magma supply from below keeps pace with the rate of sea floor spreading. Some of the magma is injected through fissures to the sea-floor where it forms basalt pillow lavas, identical in composition to the gabbro, but much finer grained because it has cooled rapidly. Magma that solidifies within these fissures forms vertical sheets, typically about one metre wide, known as dykes.

Dykes and sills

Dykes cool faster than plutonic rocks, because they are thin sheets and can lose their heat more easily. As a result, the crystals grow smaller than in a plutonic rock but coarser than in a volcanic rock. Dykes intruding other rock types are common in volcanic areas of the continents, but sheeted dykes, where dykes intrude nothing but dykes, can form only where there has been continual spreading and are known only in ophiolites and below the ocean floor.

Because they are injected up cracks or fissures, dykes record tension in the crust at the time of their formation. In many volcanic provinces dykes occur roughly parallel to one another (at right angles to the direction of crustal extension), forming a dyke swarm. Near a volcano it is common for dykes to deviate and become radial to the volcano. Volcanic eruptions that occur on the flank (side) of a volcano rather than at the summit vent are usually where the top of a radial dyke reaches the surface. Another form of dyke occurs where magma is injected up the fracture that has allowed the floor of a caldera to sub-side (Figure 5.14). This is known as a ring dyke, because of its circular plan view. Not all shallow intrusions cut up vertically like dykes. Sometimes magma is injected in a horizontal sheet, especially where the surrounding rock is composed of flat-lying beds. The generally horizontal igneous body that results is described as a **sill**. The relationship between dykes and sills is shown in Figure 6.7.

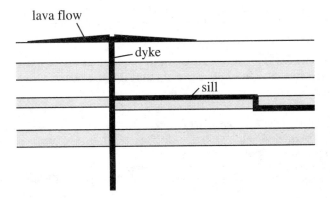

Figure 6.7 Schematic cross-section showing the difference between a dyke and a sill. Both are sheet-like, tabular, bodies. A dyke is vertical and tends to cut across the bedding of the rocks it intrudes (described as discordant intrusion). In contrast, a sill is intruded parallel to bedding (concordant intrusion), though there may be the occasional discordant step in it, as in this example.

Grain sizes in igneous rocks

We have made a generalisation in this chapter that plutonic rocks are coarse grained, shallow intrusions are medium grained. Although invariably true for plutonic rocks, there are circumstances where these rules are broken for shallow intrusions and volcanic rocks. For example, if a magma begins to crystallise by slow cooling at depth but is then erupted, it may carry with it some largish crystals that had already begun to grow. These will be surrounded by much finer crystals that did not begin to grow until the magma cooled rapidly at or near the surface. Recognition of isolated large crystals (known as **phenocrysts**) in a lava or shallow intrusive rock provides evidence that the magma had a complicated history, cooling in at least two stages.

Another exception involves the very last fraction of a per cent of magma to crystallise in a magma body. This is likely to be very rich in water and volatile gases such as fluorine, because these are preferentially retained in the magma as crystals grow. Strong concentrations of these make it difficult for new crystals to begin to form. As a result the crystals that do get started have to grow very big, in extreme cases a metre or more long, but more usually centimetres in size. Such very coarse-grained igneous rocks are described as pegmatites, and can be found in **veins** (filling irregular fractures) within igneous intrusions or the surrounding rocks.

Conversely not all volcanic rocks are crystalline at all. Magma erupted at the surface may chill so rapidly that there is no time for crystals to begin to form. The rock is therefore a frozen mixture of the elements that were present in the magma, lacking crystalline structure, and is described as a glass. Pyroclastic igneous rocks consist largely of glassy fragments, and many lavas consist of fine grained crystals floating within a glassy matrix.

Shallow intrusions, especially dykes, although generally of medium grain size (with or without a few larger phenocrysts) often have very fine crystals, or even a glassy texture at their edges. Such a chilled margin forms because the surrounding rock was cold, and the edge of the igneous body froze against it. Chilled margins are rare at the edges of plutonic intrusions because the surrounding rocks were not usually cold. There are two factors contributing to this. The first is that, being deeper down, things are warmer in general anyway. The second is that plutons are so big and contain so much heat that they heat up the surrounding rock before being cooled much themselves. This heat can cause the surrounding rock to undergo dramatic changes, by a process known as metamorphism, which is the subject of the next chapter.

7

EM

All rocks are composed of minerals. These may be crystals that grew when a magma cooled, as in the igneous rocks we considered in the previous chapters, or grains laid down to form a sedimentary rock by processes that we will look at in the next two chapters. Each mineral is stable only over a particular range of temperature and pressure. Beyond that range it will tend to break down or combine with neighbouring minerals to form new minerals.

Most igneous minerals are unstable at the low temperatures prevailing at the surface, and they tend to rot. This weathering process is slow, because, like most chemical changes, the speed of the reaction depends on temperature. However, when a rock is held at a high temperature or pressure over a long period, its constituent minerals can change dramatically to alter the character of the rock completely. Recrystallisation of this sort under the influence of temperature and/or pressure, but without melting, is known as metamorphism.

Thermal metamorphism

We closed the previous chapter with the example of a large igneous body intruding into colder rocks. When the rocks surrounding such an intrusion are examined in the field, they usually show the effects of heating by the intrusion across a zone extending a few hundred metres from the contact. This is described as contact metamorphism or thermal metamorphism.

Let's take the example of where a granite has intruded a fine-grained muddy sedimentary rock. Right against the contact with the granite we are likely to find that this mudrock is very hard and splintery, and it will have clots of new minerals that have grown within it as a result of contact metamorphism. Any splintery, spotty rock such as this is described as a **hornfels**. Prominent among the metamorphic minerals is likely to be sillimanite. This has the formula Al_2SiO_5. It grows only by metamorphism in rocks that contain aluminium and silica (a condition that is met in rocks formed from mud). As we walk away from the edge of the granite, we will eventually notice that the mudrock becomes less splintery, and the spotty pattern is lost. Soon the sillimanite disappears, and instead there is a different metamorphic mineral occurring as isolated stubby crystals up to a centimetre long. These crystals are of the mineral andalusite, whose formula is also Al_2SiO_5, but which grows at lower temperatures than sillimanite. As we continue away from the intrusion, the andalusite crystals get fewer and smaller, until we find ourselves in unmetamorphosed mudrock. We have now walked right through the 'metamorphic aureole' of the granite.

Such a metamorphic aureole is sketched in Figure 7.1. Note, however, that sillimanite and andalusite form only in rocks of the right chemical make up. In rocks of other compositions, the metamorphic minerals will be different. For example, when a granite intrudes an impure limestone, the most prominent metamorphic mineral close to the contact is likely to be the iron-rich variety of olivine Fe_2SiO_4 (maybe $Fe_{1.8}Mg_{0.2}SiO_4$), which is unknown in igneous rocks.

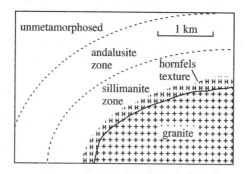

Figure 7.1 Sketch map of the metamorphic aureole of a granite intruded into mudrocks. In this example the outer metamorphic zone is characterised by the metamorphic mineral analusite, whereas in the next zone this is replaced by sillimanite. Adjacent to the granite the surrounding rocks have become brittle and developed a spotty hornfels texture.

Going back to metamorphism of rocks that were originally muddy, there is a third aluminium silicate mineral that can form if the pressure gets high enough. This is the mineral kyanite. The minerals sillimanite, andalusite and kyanite are described as polymorphs of the compound Al_2SiO_5. Figure 7.2 shows the conditions of pressure and temperature under which each will form. From this we can see that the absence of kyanite from the metamorphic aureole of the granite in Figure 7.1 means that the pressure at the time of metamorphism must have been less than about 3 kilobars, equivalent to a depth of 10 km or less. Conversely, a granite that was intruded deeper than about 15 km would have kyanite but no andalusite zone in its aureole. In addition there might be high pressure minerals of other compositions such as garnet, formula $(Fe,Mg)_3Al_2(SiO_4)_3$, which is sometimes of sufficient size and quality to be regarded as a semi-precious gemstone.

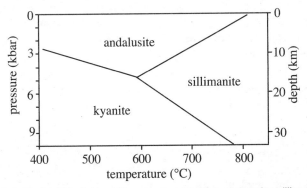

Figure 7.2 The stability fields of the metamorphic minerals kyanite, sillimanite and andalusite. These all have the formula Al_2SiO_5, but are stable over different ranges of temperature and pressure. The pressure scale is labelled in kilobars (1 kbar is equal to a thousand times atmospheric pressure), and the scale on the right gives an approximate conversion to depth in the crust.

You may wonder how it is that we can still find sillimanite surviving in the aureole of a granite exposed at the surface today, where clearly neither the pressure nor temperature is high enough for conditions to lie within sillimanite's stability field. The answer is that rocks tend to retain the minerals that grew within them at or near the conditions of highest temperature and pressure experienced during metamorphism. While a metamorphosed rock is cooling down there is usually insufficient impetus to drive the retrograde reaction from a high temperature (or high pressure) mineral to its low temperature (or low pressure) equivalent. This is fortunate, otherwise the only minerals we would ever see at the surface would be those stable at pressure of 1 atmosphere and about 0–20 °C!

Regional metamorphism

Very often pressure, rather than temperature, is the main driving force behind metamorphism. Usually the two go together, because as depth (and hence pressure) increases, so does temperature. However, it is useful to distinguish thermal metamorphism, which is caused by proximity to a hot intrusion, from regional metamorphism which is caused by the pressures and temperatures prevailing regionally. Any rock that is formed near the surface, whether volcanically or by deposition as a sediment, is liable to be buried by subsequent deposits. Eventually it may find itself at a depth where pressure and temperature are sufficient for metamorphism to begin. The whole of the lower crust consists of metamorphic rocks, except where it is made of fresh igneous intrusions.

Figure 7.3 Areas of low pressure, high temperature and high pressure, low temperature metamorphism near a destructive plate boundary.

Important departures from average conditions occur in collision zones (Figure 7.3). Where sediments have been dragged down into a subduction zone they will experience particularly high pressures, but having been recently at surface they will not be so hot as most rocks at the same depth. They will therefore experience high pressure, low temperature metamorphism, and will be characterised by appropriate assemblages of minerals. On the other hand, where the crust is heated on a regional scale by large numbers of igneous intrusions, such as below the volcanic region near a subduction zone, crustal temperatures will be higher than normal at shallow depths. The rocks here experience low pressure, high temperature metamorphism. When a paired metamorphic belt such as this is found in ancient rocks, it is a good sign that there was once a destructive plate boundary there.

Metamorphic facies

Just as in thermal metamorphism, the metamorphic minerals that grow during regional metamorphism depend on the original composition of the rock. For example, a metamorphic mineral that requires magnesium cannot develop in a rock that is deficient in this element. Because of this, rocks that have been metamorphosed under identical conditions may contain entirely different minerals. Geologists therefore classify metamorphic rocks by groupings called **facies**. Each facies reflects a different range of temperatures and pressures, and the minerals that develop in each facies differ according to the original rock type. The pressure and temperature conditions of the main metamorphic facies are shown in Figure 7.4. Metamorphism in the conditions near the top left of this diagram is described as low grade, and metamorphic grade is said to increase with distance from this point. Table 7.1 contrasts the most abundant minerals found in metamorphosed mudrocks with metamorphosed basaltic rocks in some of the facies named in this figure.

Facies	Originally mudrock	Originally basaltic rock
Greenschist	Muscovite, chlorite, quartz, sodium-rich plagioclase	Albite, epidote, chlorite
Amphibolite	Muscovite, biotite, garnet, quartz, plagioclase	Amphibole, plagioclase
Granulite	Garnet, sillimanite, plagioclase, quartz	Calcium-rich pyroxene, calcium-rich plagioclase
Eclogite	Garnet, sodium-rich pyroxene, quartz	Sodium-rich pyroxene, garnet

Table 7.1 The main minerals formed from mudrocks and basaltic rocks, metamorphosed in different metamorphic facies.

One thing that Figure 7.4 does not show is that, in the presence of water, granite crust will begin to melt at temperatures lower than those required for granulite facies metamorphism. When granulites are found, they must therefore indicate dry conditions in the lower crust.

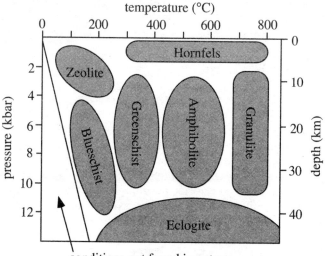

Figure 7.4 Diagram to show the pressures and temperatures required to produce metamorphic rocks of each facies. These facies grade into one another, and the boundaries between them are ill-defined, but they have been separated here for clarity.

Textures of metamorphic rocks

There is an alternative way to describe metamorphic rocks, which is simpler and more useful than facies for the geologist in the field. This is based simply on the texture of the rock, texture being a term that encompasses the size of the crystals, their shapes, and their relative orientations. We have already met hornfels, which is a textural description for a fine-grained metamorphic rock resulting from contact metamorphism, in which spotty agglomerations of metamorphic minerals have grown.

In regional metamorphism, the rocks are affected by pressure as well as temperature. When this happens, crystals tend to line up so they lie at right angles to the direction of maximum compression. This is brought about by rotation of existing crystals, and growth of new crystals in this pressure-controlled orientation. This is most noticeable for those minerals that tend to grow as flat or elongated crystals, and less apparent in minerals whose crystals tend to be more equidimensional. Alignments of minerals into planes is described as **foliation**, and such rocks are said to be foliated.

The lowest grade of regionally-metamorphosed rock is slate, which is formed by zeolite facies metamorphism of mudrocks or fine silts. Slate is well known for its ability to be split thinly along flat foliation surfaces, hence its traditional use as a roofing material. These planes of weakness, or **cleavages**, running through slate are parallel to the metamorphic foliation. As such, they have no relationship with any sedimentary layering that may have been present in the rock before it was metamorphosed. Slate is produced only from rocks that were originally fine-grained and sedimentary in nature; if a basalt or a sandstone were to be metamorphosed in the same facies as a slate it would show few obvious signs of metamorphism, except maybe under microscope examination. In particular, a coarse-grained rock cannot develop a closely-spaced slatey cleavage.

If a slate is subjected to slightly higher grades of metamorphism, still mostly in the zeolite facies, but at higher temperature or pressure, fine flakes of metamorphic mica and chlorite minerals grow on the cleavage planes. These give the rock a greenish sheen. Typically the cleavages become wrinkled, and the rock would now be described as a phyllite.

At still higher grades of metamorphism, the metamorphic minerals grow bigger, and can be identified with the naked eye. Note the different significance of crystal size in igneous and metamorphic rocks. In igneous rocks larger crystals usually reflect slower cooling, whereas in metamorphic rocks larger crystals are a result of higher temperatures and/or pressures of metamorphism.

The next grade of foliated metamorphic rock after phyllite is called schist. A schist has metamorphic minerals readily visible to the naked eye. The foliation, which is wavy rather than flat, is often made conspicuous by concentrations of mica, which grows as shiny platy crystals. Schists produced by metamorphism under normal continental conditions, where temperatures tend to rise at about 30 °C for each kilometre increase in depth, are formed under conditions of greenschist facies metamorphism, so named because the metamorphic minerals chlorite and (for rocks of basaltic origin) epidote may impart a green colour. However, under high-pressure, low-temperature conditions that define the blueschist facies, a blue amphibolite mineral (glaucophane) may develop. A schist texture can develop irrespective of whether the rock was originally fine-grained or coarsegrained.

Rocks metamorphosed under eclogite, amphibolite or granulite facies conditions are too far gone to contain abundant mica, and they rarely contain any other strongly planar minerals. In these the foliation is picked out by dark and pale minerals segregating into layers, a few millimetres thick. This type of metamorphic rock is described by the German term 'gneiss' (pronounced 'nice').

The most extreme metamorphic rocks are gneisses that reached the verge of melting. In these, veins and blobs of material crystallised from a melt are interspersed through the rock, usually strongly deformed and flattened into the plane of foliation. Such a rock is referred to as a migmatite.

The metamorphic textures described above can develop only in rocks that originally consisted of appropriate mixtures of minerals. There are two common rock types that are each made of essentially a single mineral, and so do not show these textures, under whatever facies conditions they are metamorphosed. One of these is sandstone, consisting of quartz grains. When sandstone is metamorphosed the grains tend to fuse together producing a harder rock known as quartzite, but no new minerals are formed. The other is limestone, which is a non-silicate rock consisting of calcium carbonate. When metamorphosed, limestone turns into marble, which can be a beautiful decorative stone. Recrystallised calcium carbonate usually gives it a bright white colour, and any impurities tend to concentrate into fine coloured veins.

Metamorphic rocks and most igneous rocks are products of processes within the crust. These are the hard rocks that often form the most prominent landscape features. Now we will begin to consider what happens to these rocks when they are exposed at the surface, and see how detritus derived from them accumulates to form the third main class of rocks, the sediments.

8

EROSION AND TRANSPORT

We saw in the previous chapter that each mineral is chemically stable only over a particular range of temperature and pressure. Most minerals in igneous and metamorphic rocks form at temperatures and pressures well in excess of those at the surface, so they are no longer stable by the time the rock in which they are embedded becomes exposed. However, these minerals do not vanish in a flash (otherwise we would not be able to study them!). Their conversion to new minerals that are stable at low temperatures and pressures happens very slowly indeed. In dry conditions a mineral may survive hundreds of millions of years at the surface without appreciable change. The presence of water allows minerals to 'weather' away much more quickly, although the process may be virtually imperceptible on human timescales.

Chemical breakdown of minerals during weathering

Water controls the types of minerals that develop during weathering. Large crystals of igneous or metamorphic silicate minerals tend to rot away to tiny flakes of clay minerals. The clay family of minerals is very large, but clays all have characteristics in common. They are all hydrated, having the group OH, derived from water, in their formula, and often water itself, H_2O. Their crystal structure consists of sheets of silicate loosely bound together, which is what gives clay particles their flaky nature.

The rotting of feldspar to produce a clay mineral may be described by the following chemical reaction between feldspar and water made slightly acidic by atmospheric carbon dioxide dissolved in it:

$$2KAlSi_3O_8 + 2CO_2 + 2H_2O = Al_2Si_2O_5(OH)_4 + 4SiO_2 + 2K^+ + 2HCO_3^-$$

| feldspar | carbon dioxide (in solution) | water | kaolinite (a clay mineral) | dissolved silica | dissolved potassium | dissolved bicarbonate ion |

This sort of reaction is described as hydrolysis. The clay (in this case kaolinite) remains as a solid residue, containing all the aluminium and some of the silicon that was in the feldspar, but the potassium and some of the silicon (as dissolved SiO_2) is washed away in solution.

Hydrolysis can take place both at the surface and at shallow depths where water is circulating through the rock. When carried to its extreme, and with similar reactions affecting other minerals, this can turn a granite into a deposit of pure kaolinite, which can be quarried as a source of china clay.

Hydrolysis contributes to the breakdown of micas, amphiboles and many metamorphic minerals. Potassium, sodium, calcium and magnesium tend to be removed in solution, but some can be retained in many clay minerals, for example illite $KAl_3Si_3O_{10}(OH)_2$ or montmorillonite which has the frighteningly complicated formula $(Na,Ca,Al,Fe,Mg)_8(Si_3O_{10})_3(OH)_{10}.12H_2O$.

When pyroxenes and olivines react with water, their magnesium and silica get carried away in solution, but their iron remains as a solid residue in the form of iron oxide such as the mineral hematite (Fe_2O_3).

The only common material that is more or less immune to chemical attack is quartz, which is pure SiO_2. Although SiO_2 is soluble in acidic groundwater, this is significant only for SiO_2 produced by chemical breakdown of complex silicate materials during hydrolysis. Quartz itself is virtually insoluble under surface conditions.

Dissolved calcium and the bicarbonate ion resulting from hydrolysis may combine in a new environment to produce calcium carbonate, from which limestone is made. We will look at this in the next chapter. Other dissolved elements, notably magnesium, find their way into seawater and become incorporated into minerals in the oceanic crust by means of chemical exchanges that occur below hot vents like the black smokers we met in Chapter 5. There is thus a chemical strand to the rock cycle, involving solution and re-precipitation of elements.

Although we have considered this in only barest outline, we must now turn to physical aspects of the rock cycle.

Physical weathering

Rocks at the surface suffer not only the indignity of chemical attack, they also are exposed to the abrasive action of wind, water and ice. These are very effective agents of physical weathering. For example, if water repeatedly freezes and thaws in cracks and cavities, fragments of rock can be prised away because of the expansion that occurs when water turns to ice. Even in dry conditions, a rock face may flake away because of cracking caused by expansion and contraction of the rock itself as its surface heats by day and cools by night. It may also be plucked away by the roots of plants. Fragments of rock embedded in a moving ice sheet (a glacier), or carried along by flowing water or wind, will abrade any surface they hit, and, in turn, be abraded themselves.

Where there is a steep rock face, such as a cliff undermined by the pounding of waves, lumps of rock may fall and smash under gravity. Rocks that occur in weakly-bonded layers, like slate that breaks easily along its metamorphic cleavage or sedimentary rocks that were deposited in successive beds, are particularly prone to this. So are rocks in which closely-spaced joint patterns have developed as a result of cooling (e.g. Figures 5.15 and 5.16) or release of pressure. However, the large blocks that accumulate in this way may be too big to be moved, however strong the waves or however fast the river is flowing.

For transport to occur, rocks must usually first be broken into fragments smaller than a few centimetres across. Notable exceptions occur in glaciers and landslides, which are capable of moving very much larger pieces. On the other hand, the wind rarely blows hard enough to move anything coarser than a large sand grain.

When a fine-grained rock becomes broken into transportable fragments, the particles may consist of several minerals still stuck together. Effectively they are just small pieces of rock. However, transportable fragments of a coarse-grained rock are likely to be comparable in size with the crystals within it, so in what way does a coarse-grained rock break into fragments? You might think that each mineral would become separated from its neighbours. This can happen, but a characteristic of igneous and metamorphic rocks is that the minerals are usually intergrown and interlocking, like the pieces of a three-dimensional jigsaw, so they are really rather strongly held together. Such rocks usually break across crystals, rather than between them. What tends to happen in physical and chemical weathering is that pre-existing fractures within each crystal are attacked and opened. These fractures take two forms.

First there are cracks that may have been caused by deformation of the rock, or which opened up as the confining pressure grew less when the rock was brought closer to the surface. This sort of fracture may be found in any mineral. A more important sort of fracture occurs in minerals whose crystals contain inherent planes of weakness, called cleavage planes, caused by weak bonds in its crystalline structure. This is different to cleavage in a metamorphic rock, which is planes of weakness caused by alignment of platy minerals.

Common minerals with well-developed cleavages include the micas, feldspars, amphiboles and pyroxenes. In a mica crystal, all the cleavages are parallel, so it will tend to break into flakes bounded by cleavage surfaces. The others have cleavages in two or three orientations, and break into chunkier fragments.

Most particles produced by the break-down of a coarse-grained rock therefore tend to consist of a piece of a single mineral, bounded by cleavage planes. The smaller the particle, the greater its surface area compared to its volume, and so the more vulnerable it is to chemical weathering and the shorter its survival in the transport regime before it has rotted away.

We noted earlier that quartz is effectively insoluble and immune to chemical weathering. Another peculiarity of quartz is that it is the only common mineral lacking a well-developed cleavage. Quartz is therefore physically very robust, and it survives transport very well. It should be no surprise to you that sand grains that are collected from a river bed usually turn out to be mostly quartz, unless collected from close to their source, in which case there may be a fair proportion of feldspar and rock fragments too. The other solid residue (mostly clay particles) is unlikely to be found in the same place as this material, because being much finer-grained it gets washed away to settle out eventually in much quieter conditions, as we shall see in the next chapter.

Sand-sized grains in wind and water

For now we will content ourselves with considering what happens to a sand-sized grain during transport (Figure 8.1). If broken off an igneous or metamorphic rock, a grain is likely to be quite jagged with sharp corners. This shape is described as angular. Each time a quartz grain bounces into a stationary rock, or collides with another grain, there is an opportunity for bits of the grain to be knocked off. Naturally, the corners are the most vulnerable parts. These get worn away first, so that the grain takes on a progressively more rounded shape. In contrast, a feldspar grain is more likely to break into small fragments controlled by its cleavage planes, and these fragments will rot away to clay particles, by hydrolysis.

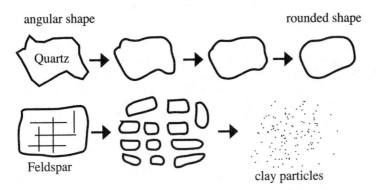

Figure 8.1 Diagram to show how a grain of quartz starts off angular in shape, but becomes progressively rounded because of collisions during prolonged transport, which abrade its corners. On the other hand, a grain of feldspar is liable to fragment along cleavage planes, and these small fragments will weather chemically leaving clay particles as the only solid residue.

Currents of water and wind are quite similar in the way they transport grains. However, water is denser and more viscous than air, so it can move a particular size of grain in a current flowing much more slowly than the speed of wind required to move the same grain. For example, water flowing at only 0.2 metres per second (less than 1 km per hour) is capable of picking up 1 mm diameter sand grains, whereas the wind speed needs to be as high as 10 metres per second to do the same. This is 36 km per hour, and corresponds to wind force 5–6 (a fresh to strong breeze) on the Beaufort Scale.

In consequence, the collisions suffered by grains transported by the wind are more violent than for water-transported grains. Sand grains blown around in a desert therefore tend to become very well rounded. Once there are no more corners to knock off, the effect of each collision is to make a microscopic pit on the surface at the point of impact, whose combined effect is to give the grain a frosted appearance. Seen in close-up under a microscope it looks just like etched glass (which is almost exactly what it is!). Another result of the high wind speed needed to transport sand grains is that rocks exposed to this 'sand blasting' may get sculpted into strange shapes.

The relatively gentle transport of grains by water means that physical weathering proceeds more slowly in water. Grains seldom become very well rounded, their surfaces do not become frosted, and obstacles do not suffer from the sand-blasting effect.

What wind and water have in common is the way in which they transport grains. Grains that are only just small enough to be moved will tend to stay close to the bed of the stream (in water) or the ground (in wind). Some simply get rolled along, but most of them bounce, suffering a collision each time they hit the surface. This mode of transport is called 'saltation', which literally means 'jumping'. Grains transported by rolling or saltation are described as belonging to the **bed-load** of the current. Smaller grains, once picked up by the flow, can remain suspended for long periods, suffering few collisions, and are said to belong to the **suspension-load**. An example might be a fast river in which sand grains are bouncing along its bed, constituting the bed-load, whereas fine clay particles are carried along in suspension, discolouring the water, and forming the suspension-load.

Moving ice

Ice is a rather different transport medium. Glaciers flow downhill at rates of between 20 and 200 metres per year, though occasionally one surges forward at over 300 metres a day. They are powerful agents of erosion, plucking rock from the sides and floor of the confining valley. They also carry with them debris that has fallen from above. Once encapsulated in the ice, these fragments cannot bounce along, no matter how big they are. Ice-transported fragments thus tend to remain angular. The only way which they can suffer abrasion is when they are carried at the base of the flow, in which case they will be scraped against the underlying rock. Debris transported by ice can often be recognised by scratches (or 'striations') on its surface, and corresponding glacial striations can be found on outcrops of bedrock that once lay below a glacier.

We have now covered the effects of erosion and transport on the material that is being transported. We will conclude our review of this part of the rock cycle by looking briefly at its effects on the landscape.

Glacial landscapes

Staying with glaciers, we have already noted their ability to pluck away at the sides and floor of the valley down which they flow. A valley that has been sculpted by a glacier can be recognised by its U-shaped profile, very different from the more V-shaped valleys carved by rivers. Another peculiarity of glaciers is that they can scour deep hollows on the valley floor, in a way quite impossible for water to achieve. If the glacier disappears, these hollows will become filled by lakes. Figure 8.2 illustrates the effects of glaciation on the landscape.

Figure 8.2 A landscape that has been sculpted by the action of glaciers. The large U-shaped valley was carved by a major glacier, which filled the valley to a depth of some hundreds of metres. Tributary glaciers fed into it, from the 'hanging valleys' on the side, whose floors are way above the floor of the main valley. The hanging valley on the left contains a mountain lake or tarn, that has filled a glacially scoured hollow. The ridges between hanging valleys are often quite sharp.

Glaciers, and more extensive ice sheets, today cover most of Antarctica and Greenland, with just an occasional ice-free peak sticking up above the level of the ice. Valley glaciers can also be found in the most of the world's high mountain ranges. However, there are large regions of the globe that are currently ice-free where the landscape bears the unmistakable signs of glacial erosion. The areas affected include North America down as far as slightly south of the Great Lakes, and Europe as far south as southern Britain. This landscape was developed during the past three million years, when the polar ice sheets repeatedly advanced Equator-ward and then retreated.

Periods of extensive ice cover are referred to as glaciations, and the warmer intervals between are called interglacials. A succession of glacials and interglacials is described as an **Ice Age**, which is a mode of the Earth's climate that has recurred for ten million years or so, roughly every 200 million years. There is no reason to think that the current Ice Age is over; we may be merely experiencing an interglacial.

Because so much water was locked up as ice during the last glaciation, global sea level was then more than a hundred metres lower than present. Many glacial valleys were cut well below present sea level, and are now flooded, to form long, steep-sided inlets of the sea, known by their Norwegian name of 'fiords'.

Among the more picturesque landscapes inherited from recent glaciations are the U-shaped lake-filled valleys of the English Lake District and the Scottish Highlands, and the fiord coastlines of Scandinavia, British Columbia and New Zealand's South Island. Among the bleaker heritage of those times are the relatively flat, glacially-scoured plains of central Canada.

Erosion by rivers and streams

Beyond the sculpting of individual rocks and boulders that we have already referred to, wind is not a sufficiently powerful erosive agent to have much effect on the landscape. In ice-free regions, it is liquid water that is the key player. Even regions with exceptionally arid climates suffer the occasional downpour. When this happens, valleys that may have been dry for a century or more become occupied by a torrential stream that may last only for a day or so.

Unlike glaciers, rivers are not nearly deep enough to fill their valleys. Therefore they erode more at the base than the sides, which explains the V-shaped profile of a typical river valley (Figure 8.3). Rivers tend to flow less straight than glaciers, so when a valley is carved by a river it is likely to be sinuous, with spurs of the hills on either side overlapping. This means you cannot often stand on the floor of a river valley and see straight along it as you can in a glacial valley. If a glacier forms in a valley previously cut by a river, it will tend to truncate the overlapping spurs, thereby straightening the valley.

Figure 8.3 A landscape sculpted by the action of the rivers and streams. If the main valley was subsequently deepened and widened by a glacier, the landscape could end up looking like that shown in Figure 8.2.

Drainage patterns

So far we have been talking in generalities. How a landscape is eroded actually depends as much on the nature of the bedrock as on the agent of erosion. This is particularly seen in the erosive effects of water. If the rock is limestone, which is soluble, rainwater may drain away through a system of underground caves, and there may be no streams at the surface at all. If the rock was laid down in a sequence of sedimentary beds, the softer units will get worn away most rapidly. If there is a strong pattern of faulting or jointing running through the rock, rivers and streams will tend to exploit and widen these lines of weakness. The nature of the bedrock, and its geological structure, can therefore often be inferred from a map showing the drainage pattern (Figures 8.4 and 8.5)

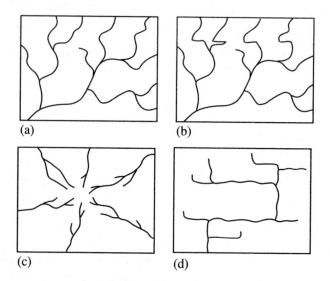

(a) (b)

(c) (d)

Figure 8.4 Different sorts of drainage pattern. (a) Dendritic, which develops where the bedrock is fairly uniform. (b) A dendritic pattern offset by recent fault motion; in this example movement on a east-west fault has offset the drainage in the northern region to the left. If the fault remains stationary for long enough the drainage pattern will readjust. (c) Radial drainage, developed over a central high point, such as a domal uplift or a volcano. (d) Rectangular drainage, which exploits planes of weakness such as joints or softer beds of rock.

Marine erosion

Finally we should look at erosion by the sea. On coastlines exposed to storms, waves can be a very powerful erosive agent, and can cut the coastline back into a cliff. Erosion is strongest near the level where waves are breaking, so it does not cut down much below the intertidal level. A wave-cut platform of planed-off rock is often exposed at low tide, extending seawards for a hundred metres or more from the foot of the cliff. There is no comparable limit to the height above sea level of the top a cliff, this being controlled solely by the height of the land surface.

Figure 8.5 Block diagram showing how the rectangular drainage pattern in Figure 8.4d can be related to the underlying geology. The structure is simple, with alternating harder (shaded) and softer (unshaded) beds dipping to the right. Streams develop along valleys where the softer rock has been worn away at the surface, and occasionally cut gorges at right angles to these through the intervening ridges of harder rock.

Where the rock is a sediment composed of gently dipping layers, a cliff is liable to be undermined, particularly if the beds at the foot of the cliff are relatively soft rock. In such circumstances it is common to find caves at the foot of the cliff. Faults or joints can be exploited too, and these often control the locations of coves and bays. As the coast retreats, a headland that once separated two bays may become a steep-sided island, or sea-stack. Some new islands of this sort inherit a cave running through them, and are known as sea-arches. Figure 8.6 illustrates successive stages of coastal erosion. Coastal retreat on this scale takes thousands to hundreds of thousands of years, depending on the strength of the rock and the violence of the storm waves.

In Figure 8.6, the coastline is likely to be marked by cliffs, except within the bays, which are sheltered from the strongest wave action and where sand is likely to accumulate. We will consider the circumstances under which sediment can be deposited, and the sort of rocks which these can eventually form, in the next chapter.

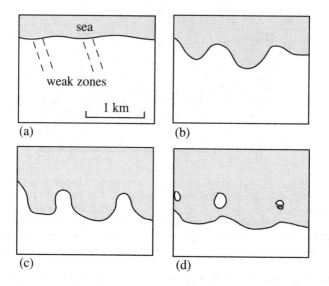

Figure 8.6 Maps to show successive stages (a–d) of coastal erosion. In (a) the coastline has weak zones running into it. These may be faults, closely spaced joints, or beds of a softer than usual rock. In (b) and (c) the weak zones have been eroded fastest, resulting in a coastline of headlands and bays. In (d) the headlands have become small islands.

9

DEPOSITION OF SEDIMENTARY ROCKS

The most hospitable landscapes, including the most profitable farmland in most countries, are usually in regions where rock is being deposited in the form of sediments, rather than being eroded. Naturally, these tend to be low-lying areas of land. A lot more sediment is deposited, unseen, below the sea.

Sediments tend to be deposited in layers, known as beds, so, after they have hardened, they form the most easily worked building stones. They are sources of certain vital resources, such as coal and oil, and may act as reservoirs for underground water supplies. Sediments can accumulate to great thicknesses, for example the Grand Canyon of the Colorado River in Arizona has cut through 2 km of sedimentary rocks, which were deposited in several episodes between about 550 and 250 million years ago.

We have already considered the processes involved in erosion and transport of detrital grains. In this chapter we will begin by looking at the conditions necessary for grains to be deposited. Many of the most familiar sedimentary rocks are formed from accumulations of detrital material. These are known as **clastic rocks**, 'clast' being a term used to describe a fragment or grain. Clastic sedimentary rocks are classified by the diameter of their most abundant clasts according to the scheme in Table Y of Appendix 2 (page 223).

Most deposits of sediment contain only a limited range of clast sizes. The reason for this is that a current of wind or water that is just weak enough to deposit clasts of a particular size will not be flowing strongly enough to bring in clasts that are much coarser than this. However, it will be flowing with sufficient speed and turbulence to keep all the significantly finer clasts in suspension.

Ripples and dunes

Except in very quiet conditions, when fine silt and mud can settle out from the suspended load, most clastic sediments are formed by deposition from the bed-load. The rolling and bouncing (saltation) modes of transport in the bed-load give rise to some distinctive features in the bed of the flow, which can very often be recognised in ancient deposits. Everyone is aware that wind tends to form sand into large dunes. Water flow produces similar but usually much smaller sedimentary structures, of which ripples are the best known. Ripples, dunes and the like are referred to collectively as **bedforms**.

Many ripples and dunes are characterised by asymmetric cross-sections. Each has a gentle slope on its upstream side and a steeper slope on its downstream, or lee, side. Grains bounce or roll up the gentle upstream slope, which is exposed to the full force of the current. As they topple over the crest they find themselves in a more sheltered region, where the current is too weak to transport them, so they simply drop onto the upper part of the lee side. This face is continually being over-steepened by the arrival of sediment at its top. Every time it gets too steep to be stable, an avalanche of grains slips down the face. Each avalanche forms a layer resting on this slip-face at the maximum angle that can be supported, which is around 30° or 40°. Grains are thus removed from the up-current side of the bedform and deposited on its lee side, and so the ripple or dune migrates slowly downstream.

Figure 9.1 shows how bedforms like ripples and dunes migrate. If there is no net deposition of sediment, all that happens is that bedforms migrate downstream. However, if more sediment arrives in the area than can be transported away, then there is net deposition. Successive bedforms migrate up the upstream faces of previous ones, and a deposit of sediment accumulates. The successive slip-faces within each bedform can often be recognised if the deposit is subsequently excavated and seen in cross-section, even after hundreds of millions of years (Figure 9.2). Slip-faces dip more steeply than the deposit as a whole, and when this is seen it is described as cross-bedding.

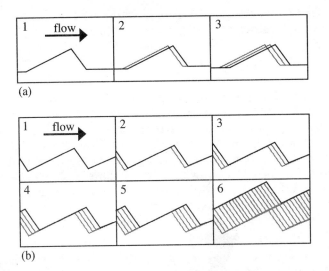

Figure 9.1 Migration of ripples (centimetres in size) or dunes (tens or hundreds of metres in size), seen in cross-section. In (a) a single ripple or dune is shown, in a situation where there is no net accumulation of sediment. The three views are at successive time intervals, with the instantaneous surface shown in black and previous surfaces in grey. In (b) the bed has a continuous array of bedforms, and there is a net accumulation of sediment so that each bedform climbs up the back of the previous one as it advances. 1–5 are successive time steps, and 6 is much later.

Figure 9.2 The face of this sand quarry in County Durham, England, shows cross-bedding formed in desert sand dunes about 250 million years ago.

When conditions are such that more sediment is being removed from an area than is being deposited, migrating bedforms may still develop, as can be seen in modern environments. However, for obvious reasons, evidence of this tends not to be preserved in the geological record.

The shapes of bedforms as seen from above depend on the conditions under which they form. A relatively slow flow in a constant direction produces transverse ripples or transverse dunes, with their crests aligned at 90° to the current. As the flow gets stronger, crests become wavier, eventually breaking into crescent shapes. These two contrasting bedforms are associated with different patterns of cross-bedding that allow them to be distinguished, as shown in Figure 9.3. Observation of the nature and orientation of cross-bedding can therefore provide very important clues as to the environment in which an ancient sediment was deposited.

There are many variations on the basic ripple and dune theme. Crescent-shaped wind-blown sand dunes are usually isolated from one another, not touching at their tips as suggested in Figure 9.3b. These are called barchans. In some areas dunes develop with their crests running parallel to the dominant wind direction; these are known as seifs or longitudinal dunes. They have slip-faces on both sides, usually attributed to corkscrew-like vortices in the wind as it is funnelled between the dunes. The largest dunes are hundreds of metres high and many kilometres in length. Clearly these are major constructs and are not going to re-orient themselves every time the wind changes direction!

Most dunes migrate at about a metre per year. The shape of a dune, and the cross-bedding it inherits within itself, is influenced not just by the annual prevailing wind but also by seasonal winds blowing from other directions. The surfaces of sand dunes are often covered in transverse ripples, and these are small enough to adjust in response to hourly variations in wind strength or direction.

Ripple-like bedforms often develop within pyroclastic flows. Here, despite the high speed of the flow, the high ratio of the pyroclastic fragments to the volume of air entrained within the flow is so great that clasts get plastered onto the ground even when the speed of the flow would otherwise be great enough to keep them moving in the bed-load.

(a)

(b)

Figure 9.3 Two sorts of cross-bedding, both produced under conditions of net accumulation of sediment. Each diagram is a block view, with flat horizontal and vertical faces cut to show how the cross-bedding appears in different planes. The actual surface, showing the shape of the bedforms, is the left-hand part of the top of each view. (a) Transverse ripples or dunes, (b) crescent-shaped ripples or dunes. The cross-bedding is different in each case: (a) has tabular cross-bedding (b) has trough cross-bedding. It would be difficult to distinguish the two if the only cross-section you saw was at right angles to the current direction, but the distinction is clear when looking along the current direction.

Turbidity currents

There are subaqueous analogues to dense pyroclastic flows, known as turbidity currents. These are particularly important for transporting sediments off the edge of the continental shelf and onto the ocean floor, notably into trenches. What happens is that a submarine landslide generates suspension of sediment. This is considerably denser than the surrounding seawater, and flows downslope as a turbulent mixture of sediment and entrained water.

Turbidity currents can be very powerful, and have been known to snap submarine telephone cables. Initially, a turbidity current is usually strong enough to scour up extra material at its base, but as the slope eases, it slows down and begins to deposit its load. At first, deposition is by plastering down of clasts as in pyroclastic flows, but as the speed decreases and the proportion of sediment remaining in the turbidity current gets less, ordinary climbing ripples form. These can be distinguished from ripples formed in ordinary water currents only by the context in which they are found. After a turbidity current has passed by, the fine-grained material that had been stirred up into suspension settles out, so the sandy rippled part of the deposit is overlain by finely laminated silts and muds. A deposit laid down by a turbidity current is called a turbidite.

Current directions

In rivers, the direction of the current is rarely as fickle as the winds, and is essentially in a uniform direction. Even though the strength of the current may vary seasonally, this uniformity in current direction controls the orientation of the cross-bedding so that unidirectional current flow can be recognised in ancient river deposits.

In the sea, tidal currents change in both strength and direction. In sediments deposited in shallow marine conditions, where tidal currents are most obvious, this is usually reflected in alternating directions of cross-bedding, and sometimes by thin layers or lenses of fine-grained suspended load that was able to settle during the period of slack water while the tide was turning (Figure 9.4).

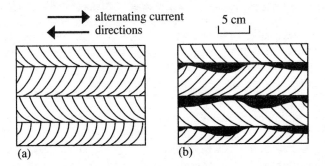

Figure 9.4 Cross-sections showing ripple cross-bedding produced by alternating tidal currents. In (a), the top of each layer of ripples has been planed off by the current bringing in the sediment that formed the next layer of ripples. In (b) there is a quiet interval between each reversal of the tide, allowing mud or silt (black) to accumulate in the hollows between ripples. This is called flaser bedding.

Waves

Figure 9.5 Cross-sections through a ripple produced in shallow water as a result of wave motion.

At sea or in lakes where water depth is less that about ten times the height of the waves, wave motion can disturb the sediment. Bedforms may grow in the shape of symmetric ripples whose opposite faces have equal steepness, quite unlike the asymmetric ripples produced by current flow (Figure 9.5). If a current is flowing in a totally different direction to wind-generated waves, current ripples and wave ripples may interfere to produce a criss-cross pattern of interference ripples. These can often be seen exposed on modern beaches at low tide. During storms, wave action may be effective at many tens of metres depth. Particularly violent storms are thought to be the cause of what is known as hummocky cross-stratification, in which hummocks form that are few metres across but only a few centimetres high, with correspondingly low-angle cross-bedding within them. When waves break on a beach, they tend to plaster sand down in successive layers, with little or no cross-bedding.

Finer and coarser sediments

Ripples and dunes are usually found only in sand or the coarsest silty sediments. Fine silt or mud will not settle out from water unless it is flowing at less than a few centimetres per second, which is too slow to sculpt sediment, however fine the grains, into bedforms. Silt and mud thus tend to accumulate in fine horizontal layers, referred to as laminations.

Conversely, when gravel and pebbles get deposited by flowing water, it is usually because the current has dropped suddenly, and there is no time for bedforms to develop. However, pebbles do tend to be oriented by the current and come to rest with their tips pointing downward into the direction the current is coming from. Each pebble thus overlaps its neighbour, in a fashion described as imbrication (Figure 9.6). This can be a useful indicator of current direction. It is virtually unknown to find an ancient deposit containing nothing but pebble-size clasts, because finer material (sand or mud) is likely to be washed in later, filtering its way down through the gaps.

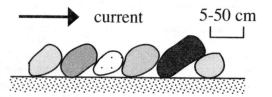

Figure 9.6 Cross-section showing imbricated pebbles, deposited by fast-flowing water. The tip of each pebble points upstream and slightly downward.

Other means of transport

We have seen previously that material may also be transported by mudflows (lahars), landslides and by moving ice. When this material gets deposited it is generally distinguishable from a water-lain deposit by the wider range of clast sizes, the angular, poorly-rounded, shapes of the clasts, and the lack of bedforms. Another distinguishing feature is that pebble-sized clasts may be supported by a matrix of mud or sand, whereas in a water-lain deposit pebbles almost always touch.

Deposition from mudflows and landslides occurs simply when the flow stops moving. Deposition from a glacier happens as the ice melts, usually at its downhill end, or 'snout'. Often, glacial detritus is re-worked by streams of meltwater, in which case it loses all or most of the characteristics listed above. Debris surviving more or less where it was dropped by a glacier, without being reworked, is referred to as moraine. When the melting snout of a glacier has retreated up a valley in stages, each successive position of the snout is marked by an accurate ridge of moraine.

Using sediments to interpret ancient environments

Most sedimentary rock is deposited under water. We have already examined the separate processes that are involved, and the bedforms that can be created. However, to interpret the environment in which an ancient deposit was laid down it is rarely sufficient to rely on such a tightly-focused approach. Rather, it is necessary to weigh up many attributes of the deposit, and to build up a picture using several lines of evidence. A rather good analogy is to consider the job of a forensic scientist, called to the scene where a dead body has been found. The coroner and jury would take a dim view if the forensic expert merely reported that the victim had been shot. They would expect to be told what sort of gun had been used, and from what angle and direction. Moreover, they would want to know how long the body had lain there before discovery, whether the body was found in the place of death or had been moved after death, whether or not there were signs of a struggle, and so on. The job of geologist interpreting a sediment is much the same.

To see how past environments can be pieced together from the evidence preserved in the sedimentary record, we will look at a few key modern depositional environments. By establishing the ranges of bedform, grain size, composition and texture, and their lateral and vertical variations and extent, we can understand what to look for in ancient deposits. This is a principle well-known among geologists, which can be summed up as the 'present is the key to the past', an expression attributed to Sir Charles Lyell (1797–1875), one of the fathers of modern geology.

Depositional environments associated with rivers

We will look first at rivers, in their sluggish downstream reaches where they are depositing rather than eroding material. Unless it has been straightened artificially, a lowland river usually winds in a lazy way across its flood plain. This flood plain consists mostly of muds and silts deposited as a result of floods, when the rate of water supply from upstream was so great that the flow could not be contained within the channel. While the flood is abating, the water covering the floodplain is sufficiently stagnant for most of the suspended load of silt and clay particles to settle out, although normally (when the river is confined to its channel) this would be flushed right down the river until it reached the sea or a lake. The flood plain is thus built of successive layers of fine- or very fine-grained flood deposits, which provide some of the globe's richest agricultural land.

Slightly more sediment is deposited on the flood plain immediately next to the river, and in this way river banks get built up, in some cases to heights of several metres. These raised banks, known as levées, are a characteristic feature of rivers such as the Mississippi.

The river itself usually flows too fast for the suspended-load to be deposited, and is likely to be transporting sand in the bed-load. We therefore find ripples when we look on the bed of a river; however these can only be preserved in the geological record if they are buried by later deposits. Burial of ripples can happen without clogging up the river, because a winding lowland river does not maintain its course. The current has to flow faster round the outside of a bend than on the inside. The river bank on the outside of a bend is therefore subject to undercutting and erosion. In contrast, bed-load tends to be deposited against the bank on the inside of a bend, where the current is slowest, producing a sand and gravel deposit known as a point-bar. The effect of this is for bends, or meanders as they are known, to become more and more exaggerated, as shown in Figure 9.7. A meander loop may eventually become cut off, allowing the main course of the river to straighten again, but another meander soon develops and the net effect over tens of thousands of years is for the river to migrate from side to side across its flood plain.

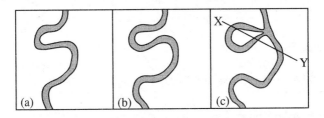

Figure 9.7 Map view at successive times showing how a meandering river migrates across its flood plain, by erosion on the outside of bends and deposition on the inside. In (c) a loop has been virtually cut-off, forming an ox-bow lake which will eventually silt up. A cross-section along the line X–Y is shown in the next Figure.

Key
 flood plain deposits (mostly mud)
abandoned channel fill (laminated mud and silt)
point-bar deposits (cross-bedded sand)

Figure 9.8 Cross-section along line X–Y in the previous Figure, showing the lateral and vertical variations in the nature of the deposits. The channels at ground level correspond to those seen on the map view. The buried channels show the course of the river at times prior to those shown in the previous Figure; these were abandoned and buried by flood-plain deposits.

The kind of thing a geologist would hope to see in cross-section, say in a cliff cut into an ancient flood plain deposit, can be pictured by constructing a cross-section across a modern environment. Figure 9.8 shows such a cross-section. Most of the deposit is muds and fine silts, laid down on the surface of the flood plain. Abandoned channels become filled by laminated silts and muds, looking rather similar to flood plain deposits except they will have steep edges marking the sides of the channel where it was cut into the flood plain. Successive positions of active channels are marked by point-bar accumulations of sand and gravel deposited in the inside of bends. The extent of each such sand/gravel sheet depends on how far the bend migrated before the channel was abandoned. Within each point-bar deposit, the grain-size gets finer upwards, with coarse sand or gravel at deeper levels, representing a faster current, and finer sand towards the top where the current was slowest. Ripple cross-bedding indicates the direction of the current.

Sediments laid down by a meandering river thus consist mostly of very fine-grained material (mud and silt), most of which was laid down on the flood plain. However, there are sheet-like or lens-like bodies of gravel and sand within it that provide the evidence for migrating channels.

Not all rivers have this sort of flood plain. Downstream from the ends of glaciers, or in regions where rainfall is strongly seasonal, rivers have to carry a particularly high sediment load at certain times of the year. Their downstream reaches are not confined within a single channel. Instead the river occupies several interconnecting channels, separated by bars of gravel that are covered only when the river is in flood. This is known as a braided river (Figure 9.9). Looked at in cross-section, the deposits from a braided river consist almost entirely of lenses of sand and gravel. Any mud is restricted to the occasional abandoned channel, so it is relatively easy to distinguish between deposits from meandering and braided rivers.

Figure 9.9 Map view of a braided river, which may cover an area of a one hundred metres to about ten kilometres in width. Braided rivers are common in Texas (the Brazos is a classic example) and other areas where discharge is strongly seasonal.

Flash floods

The mountainous fringes of desert areas experience even more strongly seasonal river flow. Here, channels may be dry for almost the entire year, or even for centuries at a time. However, when it does rain it usually does so very heavily, and gullies that are almost always dry may rapidly become brim-full with water flowing fast enough to roll even quite large boulders along in the bed-load. A flash-flood such as this is one of the chief hazards to beware of when working in mountainous terrain. Where a gulley debouches onto the plains, the torrent is no longer confined and so it disperses. This causes the speed of flow to drop dramatically, so the larger clasts in the bed-load come to rest. The deposit spreads out radially from the mouth of the gully, to form an alluvial fan. Seen in cross-section, it consists of successive layers of imbricated boulders and pebbles (with sand washed in between later). Channels within it may not be obvious, but, with care, erosion surfaces can usually be distinguished at the base of each fresh influx of sediment.

Deltas and estuaries

Turning our attention now to where rivers reach the sea, we often find a delta built up of sediment deposited by the river. There are two factors contributing to this. The first is that, just as in an alluvial fan, where the flow spreads out, its speed drops, so any bed-load that the river has managed to bring all the way to the sea will now be deposited. The second is a result of something that happens to clay particles as soon as the water becomes salty. Clay particles are tiny, and remain in suspension provided they stay as individual particles. However, once they get into even slightly salty water they start to stick together, or flocculate. The clusters of clay particles created in this way are big enough to sink, and so the suspended load of the river gets dumped as soon as the river water mixes with the sea water.

The world's two most famous deltas show some important differences in morphology, resulting from differing local conditions. The archetypal delta is that of the Nile. The resemblance of this on a map to the shape of the Greek capital letter delta, Δ, led the geographer Herodotus to coin the term 'delta' to describe it, in 490 BC. The Nile breaks into a plethora of distributory channels, radiating outward from the vicinity of Cairo, more than a hundred kilometres from the sea. The seaward fringe of the delta is a gentle arc, rather than the straight edge implied by the Greek character. In contrast, the Mississippi delta has been likened to a bird's foot in appearance, because the levée-confined channel that feeds it has built out beyond the main shoreline and it and the few main distributory channels coming from it stick out into the sea like the toes on a foot.

The main reason for the different shapes of these two deltas is that the Mississippi discharges much more sediment than the Nile. It is also less exposed to strong waves, which tend to mould a delta front into a Nile-style arc.

Neither the Nile delta nor the Mississippi delta is subject to strong tides. Where there is a large tidal range, for example at the mouth of the Ganges-Brahmaputra, tidal currents scouring up and down the distributary channels flush a lot of the sediment seawards, and keep the channels wide. Other river mouths lack any kind of delta, but are funnel-shaped inlets of the sea known as estuaries. Most of the river mouths around the British Isles and on the east coast of North America from Chesapeake Bay northward are of this kind. Factors encouraging the presence of an estuary rather than a delta include relatively low amounts of sediment discharge and high tidal ranges. Some estuaries are valleys that have been partly submerged by the rise in sea level that happened as a result of the melting of the main ice sheets at the close of the most recent glaciation.

In many ways the tops of deltas resemble flood plains of meandering rivers. The deposits found there consist of ripple cross-bedded sand that accumulated on point-bars, silts and muds laid down in abandoned channels, and mud deposited over the delta plain during floods. Delta plains are often densely colonised by vegetation, and roots, fallen leaves and branches can often be found as fossils. Decaying vegetation may accumulate thickly enough to form coal deposits, as we will see in Chapter 11.

Moving seawards from a modern delta, the sandy sediment deposited in the actual shoreline may be reworked by wave action, which is reflected in the style of cross-bedding. This area is referred to as the delta bar; channels are largely absent here, and so is mud because such shallow water is usually too agitated for clay to settle, even when flocculated. Another diagnostic feature is broken or complete shells of marine organisms, which are, of course, absent in any sediment deposited higher up the river. Going slightly further offshore, to the delta front, we reach the zone where most of the flocculated clay is able to settle. This region is characterised by unbroken shells of marine organisms, and burrows in which these and soft-bodied organisms lived.

Prograding deltas

We can now introduce a very important concept. By their very nature, deltas tend to build out seawards over time. The shoreline is said to be **prograding**. As a result, if we follow events at a specific point, originally a few kilometres offshore, over maybe 10 000 years we might record the following sequence. Purely marine deposits are encroached over by delta front sediments. At first these will be the very finest clay material, which has been washed furthest from the delta, but as the delta approaches coarser clay and silt are deposited here. Throughout this time the sea bed is getting shallower, because of the accumulation of sediment. Eventually we will find ourselves on the shoreline, where the sands of the delta bar are being deposited. The shoreline builds out seawards, and we find ourselves in a distributary channel of the delta, whose base may erode away some of the underlying delta bar sands. After a while our channel migrates away, and we find ourselves on the delta plain, though there is always a chance that the same or another channel may migrate our way if we wait long enough.

Those without the patience to wait 10000 years could see the same variety of environments by moving sideways a distance of ten or so kilometres from offshore to delta top. We could also see the same sequence of environmental changes by drilling a 100-metre-long, vertical borehole though the top of a modern delta. This would reveal delta plain sediments at the top, delta bar sediments below, and progressively finer delta front sediments below these (Figure 9.10). In other words, the lateral juxtaposition of sedimentary environments at an instant in time is reflected by environmental changes seen in a vertical sequence. This relationship is known to geologists as Walther's Law, after Johannes Walther who first stated it in 1894.

Figure 9.10 shows the delta plain sediments overlain by offshore marine sediments. This might surprise you, because continuing our imaginary sojourn on a delta, you might expect that eventually we would find ourselves well and truly on dry land as the shoreline continues to prograde. This is the long term tendency, but it is quite likely that we will see several returns to marine conditions in the meantime. This is because if our river bursts through one of its levées near the upstream part of the delta, the main discharge may be diverted into a new channel, in which case a new area of delta will build out into an adjacent area of the sea. This will starve our observation point of sediment, which will rapidly sink below sea level as a result of compaction of the underlying sediments, and downward displacement of the crust as a whole because of the weight of sediment.

Offshore marine

Coal
Delta plain laminated silts and muds
Distributary channels, point bar sands

Delta bar, sands reworked by waves

Delta front, laminated sand, silt and mud. Shells and burrows of marine organisms becoming common towards the base

Offshore marine

Figure 9.10 Changing grain size and bedforms seen in a vertical sequence through a prograding delta. The column is drawn with variable width corresponding to changes in grain size. The basic sequence, fine-grained (marine) at the bottom, getting gradually coarser upwards and topped by non-marine muds is characteristic of deltas in general, although the details depend on the relative contributions of waves, tides, river flow and sediment supply.

To take the example of the intensively studied Mississippi delta, we know that the present-day main channel, flowing south-eastwards to the sea from New Orleans, came into existence only about 1000 years ago. For most of the previous 5000 years, the main discharge went southwards from an area near Baton Rouge to produce a (by now mostly sunken) delta to the west of the present one. The same period also saw an interval when the main discharge was eastwards from

New Orleans, producing a delta to the east of the present one, part of which is still above sea level.

Logging a sedimentary sequence

Geologists rarely have the luxury of cliffs that are high enough and continuous enough, and running in the correct orientation to record the whole lateral extent of a delta. However, by interpreting the environmental changes seen in a single quarry face, or in a core recovered from a borehole, or in a series of isolated outcrops, they can construct a log such as that in Figure 9.10 and then use Walther's Law to deduce what the whole area would have looked at the time when the sediments were being deposited.

Beaches and other coastal environments

The concept is not just applicable to deltas; it works whenever there have been gradual changes in sedimentary environment. For example, a sedimentary sequence deposited over time at a place where the shoreline was prograding by the accumulation of sand washed onto a beach would begin at its base with marine deposits characteristic of deepish water, below the level exposed at low tide, and too deep for storm waves to disturb. These sediments, deposited in conditions of unvarying low energy, would be very fine grained, probably either mud or limestone (which we will discuss shortly). The slow accumulation of this fine-grained deposit would eventually bring the sea floor to a shallow enough depth for storm waves to disturb it, and we might expect to see erosion surfaces and sandy layers, possibly containing hummocky cross-stratification representing big storms.

As the sea floor became shallower and the general environment conditions became more energetic, with stronger tidal currents and disturbance by waves at every low tide, mud or limestone would be replaced by silts and fine sands. Such conditions are favoured by many burrowing organisms that feed off the continual supply of nutrients brought in by the currents. Cross-bedding, such as wave-generated ripples, is often disturbed or obliterated by burrows. We might also find complete fossilised shells or organisms, preserved within their burrows.

Continuing up the sequence we would reach sand deposited on the beach, between the low and high tide marks. This is generally laid down in low-angle sheets with occasional ripple cross-bedding. There would probably still be plenty of shelly material, but now it would be mostly broken fragments. If our beach was the sort with sand dunes behind it, the top of our sequence would contain the large-scale cross-bedding that is characteristic of wind-blown dunes. Because the sand had not been blown very far, the grains would be unlikely to have become very well rounded and frosted, as it is typical of desert sand dunes. Eventually, beach dunes get colonised by plants, and so we might find a few rootlets in our deposit.

The sequence we have described would be most likely to be preserved in the geological record if it sank below the sea level and was buried below another cycle of prograding shoreline sediments. Otherwise, it might be eroded away by river action, or by a change in marine conditions, For example, if rather than bringing a net supply of sand to the beach, the currents started to take more sand away than they brought in, the beach and eventually the dunes behind it would be washed away.

There are many variations on the theme of coastal sediment deposition. If the tidal range is great and the gradient gentle, waves break a long way offshore, and have lost their energy by the time they lap up to the high tide mark. Particularly on sheltered coastlines, there may be a sand beach exposed at mid-tide, backed by an extensive mudflat consisting of very fine-grained material that settles out in quiet, wave-free, conditions while the tide is turning. Today in temperate regions, these tidal flats become colonised by coarse grasses to form salt marshes. The grass roots bind the sediment together, preventing it from washing away and encouraging it to build up above the normal high tide level. In humid tropical regions, the same function is served by mangrove trees, whose roots spread out above ground level and act as a trap for sediment.

Evaporite deposits

In arid regions, there is insufficient rainfall to support salt marsh or mangrove swamp vegetation. Instead, above the normal high tide mark mud is matted down by layers of salt-tolerant fibrous algae. Such a salt flat is known by its Arabic name of 'sabkha'. The mud here is not dominated by clay minerals, but by mud-sized grains of carbonates, chiefly calcite, $CaCO_3$, and dolomite, $CaMg(CO_3)_2$.

There are organic ways of growing these carbonates that we will look at shortly. However, on a sabkha they form because of an inorganic process. The sun beating down on the surface of the sabkha causes evaporation, and seawater is continually drawn in through the mud to replace what is lost. As the water evaporates, its dissolved constituents come out of solution. Minerals formed in this way are described as **evaporites**. If the rate of evaporation is sufficient, other evaporite minerals begin to grow, notably gypsum, $CaSO_4.2H_2O$, anhydrite, $CaSO_4$, and halite, $NaCl$, which is common salt. Gypsum deposits make alabaster, much used by sculptors and from which plaster of Paris is made. Common salt, of course, has many uses.

Evaporite minerals do not only form in sabkhas. They can also form in the sea, where evaporation of seawater is proceeding faster than

Figure 9.11 These large 'swallow tail' crystals of the evaporite mineral gypsum, today exposed on the uplifted island of Cyprus, grew on the floor of the dessicating Mediterranean Sea about five million years ago.

the rate of replenishment by fresh water. Such conditions are met in enclosed marine basins in sub-tropical latitudes where there is little rainfall. Here, the sea bed becomes covered by crystals of evaporite minerals that have grown near the surface, where the effects of evaporation are most extreme, and sunk to the bottom. There are no extensive marine evaporite deposits forming today, but there is up to two kilometres of evaporites on the floor of the Mediterranean, deposited between 5.4 and 4.8 million years ago when the Mediterranean was periodically cut off from the Atlantic by the emergence of the Straits of Gibraltar above sea-level (Figure 9.11). To produce this thickness of evaporites would require the Mediterranean to have been completely evaporated about forty times. Even today, the rate of surface evaporation from the surface of the Mediterranean Sea exceeds the rainfall and river influx to it, and it is only replenishment by Atlantic water that stops it drying out, which would take only about a thousand years if the Straits of Gibraltar were closed.

Economically important evaporite minerals that formed in an enclosed marine basin known as the Zechstein Sea about 220 million years ago are found in a deposit stretching from north-east England to northern Germany. These include potassium salts, and are the foundation of the British chemical industries of Teesside.

Limestone

Evaporites form only under special circumstances. However, calcium carbonate is the main rock-forming mineral in many sedimentary environments. It owes its ubiquity to biological, rather than physical or chemical processes. A rock made mostly of calcium carbonate is called limestone, which is by far the most abundant non-silicate rock type. Deposition of calcium carbonate is an important part of the rock cycle, in which calcium that was dissolved during weathering and the bicarbonate by-product of hydrolysis come out of solution maybe thousands of kilometres from where the calcium was dissolved. Calcium carbonate is most commonly found as the mineral calcite.

Sea shells found on the beach are made of calcium carbonate, and some limestones are made of deposits of such shells, or of fragments of such shells. More important globally are microscopic, generally single-celled, plants and animals that live in the sea. Most drift around near the sea surface where there is plenty of sunlight. These are referred to as plankton. Less abundant forms live on the sea floor. The most important animal forms are foraminiferans, which are simple organisms similar to the amoeba except for their often multi-chambered calcareous shells, commonly up to about a millimetre across. The most important plant variety is the coccolithophores.

These are smaller and rather than having a single shell they are covered in an array of disc-like plates called coccoliths, typically only 20 micrometres in diameter.

When such organisms die, their remains sink towards the sea floor. Provided conditions are gentle enough, they can settle to form extensive deposits. The famous White Cliffs of Dover are cut into a 70-million-year-old accumulation of coccoliths that was deposited on the floor of a calm, shallow sea that had no major rivers discharging muddy sediments into it (otherwise the cliffs would not be so white). This rather special kind of limestone is called chalk.

Deposits of fine-grained calcium carbonate derived from the shells of marine plankton cover much of the deep ocean floor in places far from land where there is little other sediment supply. It is thought the removal of carbon dioxide into limestone through the action of marine plankton is a natural mechanism that maintains the concentration of carbon dioxide in the atmosphere at an equitable level. This may help to soak up much of the carbon dioxide released into the atmosphere through burning of fossil fuels, and so act as a brake on the greenhouse effect.

Tiny shells dissolve slowly as they sink, especially in cold waters, and in the deepest and coldest regions deep ocean sediments tend to be dominated by the microscopic shells of planktonic plants called radiolaria that secrete silica rather then calcium carbonate.

In some environments, limestone can form as a ready-made hard rock rather than an unconsolidated deposit. This occurs where the sea floor is encrusted by sedentary organisms that secrete calcium carbonate. Today, the most important such encrusting organisms are corals. These are animals but have photosynthetic algae living symbiotically within them, so they require clear shallow water with sufficient sunlight. Delicate branching or fan-like species grow in quiet conditions, and more robust species colonise rougher settings.

Coral reefs are familiar to most people through marine wildlife television documentaries. If the substrate they are built on is subsiding, they can build upwards so as to keep pace and thereby maintain the living part of the reef close to sea level. This was realised by Charles Darwin in his explanation for coral atolls. which are rings of reef made of robust, wave-resistant coral sheltering a shallow lagoon that hosts more delicate corals (Figure 9.12). Drilling has shown that the volcanic basement to some atolls has sunk to more than a thousand metres below sea level. Mururoa atoll in the Pacific, which achieved notoriety in the mid-1990s as the site of the French underground nuclear testing, is of this type.

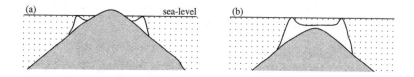

Figure 9.12 Cross-section to show two stages in the formation of a coral atoll by subsidence. An extinct volcanic island is fringed by a reef. The island sinks isostatically (in response to the load exerted on the lithosphere by the weight of the island), but the reef builds up to keep pace with it.

Apart from forming atolls, coral reefs shelter many tropical shorelines in places devoid of muddy river sediment. The biggest and most famous of these is the 2000-kilometre-long Great Barrier Reef off the coast of north-east Australia.

Along some carbonate-rich coastlines, agitation by waves causes a rolling motion that can cause near spherical millimetre-sized grains of calcium carbonate to form, by concentric growth around a tiny fragment or silt particle. These grains are called ooids, and they can accumulate into deposits forming rocks called oolites, or oolitic limestone.

Turning sediment into rock

It has been possible to give only a general account of the more distinctive sedimentary environments. We must finish by considering what it takes to turn a deposit of sediment into a rock.

The main requirement is that it must remain buried, and not be eroded away. It is easy to forget this obvious point when looking at a sedimentary rock in the field. A three-centimetre-thick layer of ripple cross-bedded sand may have taken less than a day to accumulate, but it may be the sole survivor of a million such layers that were successively deposited and then stripped away, perhaps with every turn of the tide. Such is the fate of most sediment deposited near shorelines, in rivers and in desert dunes, and it is only in deep marine conditions that we normally find continuous succession without erosional breaks.

Once buried, processes begin that turn a soft sediment into a hard sedimentary rock. Only reef limestones and some evaporites are formed in an initially consolidated fashion, other sediments begin as loose, unconsolidated deposits. However, with burial, and over time, various chemical and physical changes take place, collectively described as **diagenesis.**

The dominant physical process is compaction, caused by the weight of later sediment deposited on top. Clay-rich muds may begin with solids making up only about 20% of the volume, the remaining 80% being spaces between clay flakes filled by water. The deposit is said to have a porosity of 80%. By the time such a deposit is buried by a kilometre or so of sediment, most of this pore water has been squeezed out, and the porosity is reduced to less than 10% (Figure 9.13). A one-metre layer of compacted mudrock may thus represent several metres of uncompacted mud.

Figure 9.13 Magnified vertical cross-section to show compaction of mudrock. (a) Clay flakes oriented randomly soon after deposition. (b) Clay flakes rotated during compaction to lie horizontally; in this state the rock can be split easily along closely-spaced horizontal planes and is so called a shale.

Sands and silts are made of more equidimensional clasts than clay, so the potential for compaction and loss of porosity is less (Figure 9.14). Compaction is achieved by a combination of rotation, bending and fracturing of grains. A sandy sediment may begin with up to 40% porosity and end up with only 10% porosity at depth of about a kilometre. At greater depths, the points of grains resting on others may dissolve away through a process called pressure solution, reducing porosity still further. Similar compaction processes affect carbonate sediments.

Figure 9.14 Magnified vertical cross-section to show compaction of sandstone. (a) Grains soon after deposition. Each grain touches its neighbours, but only at one point, which rarely lies exactly within the plane of this cross-section. (b) After compaction. There has been some rotation of grains, but in this case most of the compaction has been achieved by pressure solution, in which material has been dissolved from points where grains touch, This has been reprecipitated nearby, so that most of the remaining pore space is filled by cement.

Most of the chemical changes during diagenesis are controlled by the nature of the pore waters. Silica that has been dissolved by pressure solution may come out of solution to fill the remaining pore spaces with silica **cement**. This can turn a loose sandy deposit into a very hard rock in which the grains are as thoroughly stuck together as crystals in an igneous or metamorphic rock. Pressure solution in limestones results in pore waters that are saturated in calcium carbonate, and calcite cements are almost universal within limestones. Calcite cements may also form in the pore spaces of other rock types, either from solution of minor amounts of calcium carbonate originally in the deposit, or from calcium carbonate carried in from elsewhere by migrating pore waters.

There are some mineralogical changes that may take place too. One of the earliest of these affects limestones. In many shells, the calcium carbonate takes the form of aragonite, which has the same formula as calcite but a different crystalline structure. Fragmental debris created from the abrasion of shells is therefore mostly aragonite, but this converts to calcite during diagenesis, so hard rocks made of aragonite are unknown. Sometimes the diagenesis of limestones from originally lime-rich muds involves recrystallisation so that the resulting rock consists largely of interlocking crystals of calcite. Its texture may resemble an igneous rock rather than a fragmental sediment, but its composition and the common presence of fossils makes misidentification unlikely.

Other diagenetic changes include the driving off, as a result of pressure and mild heat, of water that was originally bound up in clay minerals. New clay minerals, with less water in their formulae, take their place. This sort of change is on the verge of metamorphism, and it is really just a matter of choice what to describe as extreme diagenesis and what low grade metamorphism.

We have now seen how sedimentary rocks are created from components derived by weathering and erosion. Sedimentary rocks are not a dead-end in the rock cycle. Most ocean floor sediments eventually get dragged into trenches at destructive plate margins, where they are metamorphosed or even melted. Sediments in shallow seas or deposited on land may be regionally metamorphosed if they are buried to sufficient depth, or if they get caught up in a continent–continent collision. Alternatively they may be raised up to be exposed to the forces of weathering and erosion. The remaining links in the rock cycle are covered in the next chapter, where we see how rocks behave when subjected to stresses that deform them.

10

DEFORMATION OF ROCKS

Parts of the Earth's crust are from time to time subjected to forces that tend to extend them or compress them. These forces may be regional in extent, such as those associated with plate tectonics, or they may be more local, perhaps caused by the extra load on the crust produced by a thick accumulation of sediment or the growth of a volcano.

When the temperature is high, rock tends to deform in a ductile fashion, that is to say it can be squeezed or stretched without fracturing. This occurs by deformation, rotation and recrystallisation of grains or crystals within the rock, and happens most readily if the deforming forces build up slowly.

Ductile deformation is characteristic of deformation at deep crustal levels, where the temperature is higher. At shallower levels, rocks tend to fracture, especially if forces build up quickly. This is how faults form, and is described as brittle deformation.

The effects of deformation are most readily seen in sedimentary rocks, because their inherently bedded nature provides ready-made markers. For example, an individual bed of limestone, sandstone or mudrock may be traceable over tens or even hundreds of kilometres. If its base and top were virtually horizontal when deposited (provided we are sure we are looking at the real top and bottom boundaries of the bed, rather than cross-bedding within it), and if the bed is no longer horizontal when we see it then it must have been tilted. Similarly, if the bed is no longer flat but curved, then it must have been bent. Moreover, some beds within a sedimentary sequence are usually distinctive, and enable us to match displacements across faults.

Plate 1: Vent inside the active crater of Masaya volcano, Nicaragua, filled with molten lava at about 1100 °C, seen by night. The vent is several metres across. A few seconds prior to this photograph a small explosion distributed blobs of lava beyond the rim of the vent; these are beginning to cool down, and so are glowing less strongly.

Plate 2: Explosive eruption of Mount Etna volcano. A sustained fountaining event is sending incandescent material to a height of about a hundred metres above the rim of the crater, and an explosion has produced a smaller cloud of grey ash directed obliquely towards the viewpoint.

Plate 3: Ropey (pahoehoe) lava on the flanks of Kilauea volcano, Hawaii.

Plate 4: Vertical sheets of basalt feeding the overlying pillow lavas, in the Troodos ophiolite of Cyprus. This region was created by sea-floor spreading on the ocean floor. The major dyke on the right is about a metre wide.

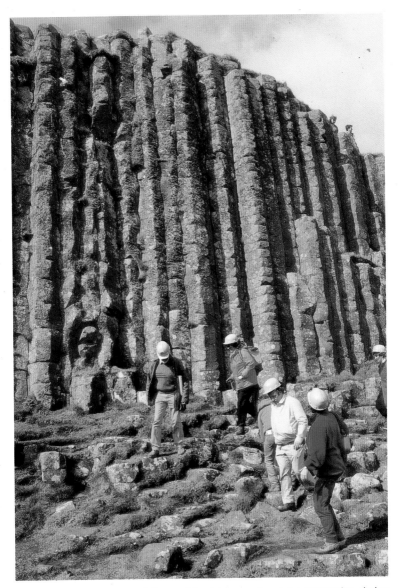

Plate 5: Vertical columns delineated by joints produced by contraction during cooling of a thick basaltic lava flow. (Giant's Causeway, County Antrim, Northern Ireland)

Plate 6: Close-up view of a granite. Pink and white crystals are two varieties of feldspar mineral, grey crystals are quartz, and black flecks are biotite. The brassy metallic minerals are sulfides of copper (pyrites and chalcopyrites) and the bluish metallic mineral is molybdenum sulfide (molybdenite). (Shap, England)

Plate 7: Cross-bedding in sandstones deposited in a braided river, approximately 300 million years ago. The coin provides a scale. (Roaches Grit, Staffordshire, England)

Plate 8: The upper two-thirds of this cliff shows cross-sections through a series of river channels, each containing the characteristic cross-bedding of point-bar deposits. (Kalbarri, Western Australia)

Plate 9: This deposit of small boulders was dumped here when the force of a flash-flood was spent. (Wadi Ahin, Sultanate of Oman)

Plate 10: Complexly deformed gneiss, in which the original metamorphic layering has been folded and then re-folded to produce the complex interference pattern visible on this surface. The pale quartz-filled vein to the left of the hammer is a later feature. (Yilgarn block, Western Australia)

Plate 11: An open-cast copper mine, whose ore is a result of hot fluids circulating through the ocean crust near a spreading axis. The pit is over a kilometre across. (Skouriotissa, Cyprus)

Plate 12: A 10 cm long piece of limestone showing abundant internal and external moulds of bits of crinoid stem. (Cromford, Derbyshire, England)

Plate 13: Satellite image covering an area about 30 km across within a granite-greenstone terrain. Approximately the same area is shown in Figure 13.2 (Pilbara Block, Western Australia)

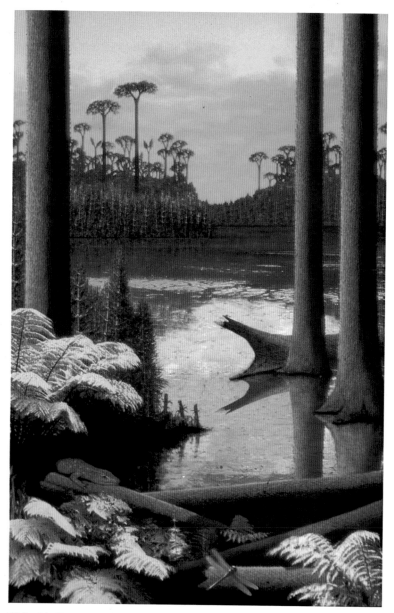

Plate 14: Reconstruction of the swampy conditions that led to the formation of coal in Carboniferous times. The vegetation is of types known mainly from fossils. The dragonfly has a wingspan of about 60 cm.

Different types of steep fault

We will start by looking at brittle deformation, or faulting. There are essentially three types of fault movement, all illustrated in Figure 10.1. The rock all around has been removed to enable us to see what happens to a specific block of terrain. The sequence here consists of an undeformed succession of sedimentary beds cut by a single dyke of igneous rock, which will act as a convenient vertical marker. For our purposes, the sedimentary beds could equally well be lava flows, layers of pyroclastic rocks or a series of horizontal sills; the main thing is that there is a set of individually recognisable horizontal layers. A useful term that encompasses all kinds of layers, not just sedimentary beds, is strata.

In Figure 10.1 (a) the area is shown before any fault motion has taken place, though the position of the future fault is marked. In this example it is a plane that is not quite vertical but is dipping steeply down to the right.

One thing that might happen is that the blocks of terrain on either side of the fault might slip sideways past each other. This is called strike-slip or **transcurrent fault** motion, and is what is happening along most of the San Andreas fault in California. As Figure 10.1 (b) shows, if the direction of motion is exactly parallel to the bedding, there is no mis-match of strata either side of the fault. However, the vertical marker, in this case a dyke, has been displaced, and the amount of offset can be measured to determine how much movement there has been across the fault plane since the dyke was intruded. If we were to stand on the dyke on one side of the fault, we would see that the other part of the dyke had been offset to the left. We would see a leftwards offset whichever side of the fault we were looking from. The fault illustrated is thus distinguished as left-lateral (or sinistral) transcurrent fault. If the offset had been in the opposite sense, so that the dyke had been displaced to the right, it would be called a right-lateral (or dextral) transcurrent fault.

Alternatively, the fault plane could separate regions whose relative movement is vertical. It is immaterial whether we regard one side as going up or the other side as going down. When this happens, the offset can be measured by the fact that strata that were originally at different depths have become juxtaposed. However, there will be no evidence of offset displayed by a vertical feature displaced by the fault. Figures

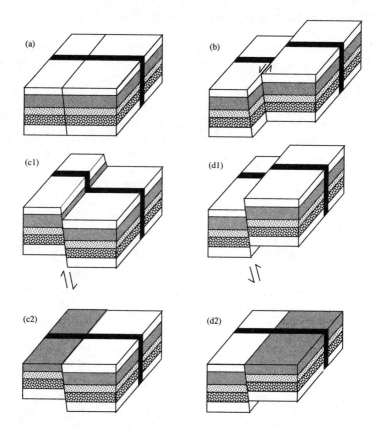

Figure 10.1 Cross-sectional block diagrams to illustrate three types of fault, each starting from the situation in (a). (b) Shows the area after transcurrent movement across the fault plane. (c) Shows what would happen instead if the block on the right-hand side went down (c1 with no erosion, c2 after erosion), and (d) shows what would happen if the block on the right-hand side went up instead. Note that the fault plane is not quite vertical, we can therefore distinguish case (c) as a normal fault and case (d) as a reverse fault. See text for further discussion.

10.1 (c) and (d) show alternatives, each starting from the situation shown in Figure 10.1 (a). In Figure 10.1 (c) the terrain on the right has gone down relative to that on the left. Usually, the side that is being relatively uplifted experiences much more erosion, and the result of planing off to new horizontal surface is shown in Figure 10.1 (c2). As a result, an older bed on the left of the fault is exposed at the

surface next to a younger bed on the right of the fault. In Figure 10.1
(d) the fault motion has been in the opposite direction, and the
younger bed survives at the surface on the left of the fault instead.

Figures 10.1 (c) and (d) are not quite mirror images of each other,
because the fault plane is dipping steeply downwards to the right
instead of being vertical. In Figure 10.1 (c) the downthrown terrain is
on the down-dip side of the fault; this is the commonest kind of fault
movement, and is known as a **normal fault**. In Figure 10.1 (d) the
downthrown terrain is on the up-dip side of the fault. This is
described as a **reverse fault**. In active fault belts, erosion may keep
pace with the rate of uplift, so that faults are rarely represented by
cliffs. There is certainly never an overhang at the surface of the kind
shown in Figure 10.1 (d), because this would crumble away as soon as
it formed. However, sometimes erosion lags behind to the extent that
the position of an active normal or reverse fault is indicated by a
straight edge to a belt of hills.

If you compare stage (c) in Figure 10.1 with stage (a), you will see
that because the fault plane is not quite vertical, the distance from
left-to-right has increased slightly. Normal faulting is common in
regions of crustal extension. If however, the fault acts as a reverse
fault, ending up at stage (d), the left-to-right distance has been
reduced. Reverse faults are characteristic of regions of crustal compression. However, these are gross generalisations. In nature most
faults are parts of linked systems and it is unwise to read much significance into a single fault.

A series of consecutively developed normal faults appeared in Figure
4.6, which showed continental rifting. Figure 10.2 shows a similar situation at an enlarged scale, to define some terms that you are likely

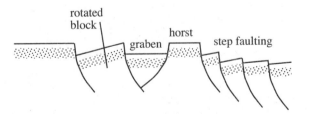

Figure 10.2 Cross-section to show some effects of faulting. A graben is a block that
has dropped down between two normal faults. A horst is a block that has been left
standing high by blocks on either side dropping down on normal faults. In nature, the
scale of this would typically be kilometres to hundreds of kilometres across.

to find used to describe faulted areas. Note that the faults in this figure are shown as curved surfaces, becoming less-steep at depth, which allows rotational movement between adjacent blocks.

Low–angle faults

The fault plane drawn in Figure 10.1 is very steep, and cuts the bedding at a high angle. Sometimes fault planes lie at a low angle to bedding, in other words nearly parallel to it. In this case, where the motion is compressional the fault is described as a thrust fault, or more commonly just as a **thrust**. Where the motion is extensional, the fault is described as a low-angle normal fault or low-angle extensional fault. Faults that develop parallel to bedding do so because some layers are easier to fracture than others, and thrusts and low-angle normal faults usually lie between beds or within weak beds such as shales, and avoid stronger beds such as well-cemented sandstones and limestones. However, they must approach the surface somehow, and this is usually achieved by the fault plane cutting steeply through each strong bed yet dipping gently everywhere else. In this situation it is more appropriate to write of a fault surface rather than a fault plane.

Figure 10.3 is a rather complicated block diagram, but it serves to illustrate many of the most important features of low-angle faults. A thrust is illustrated, but the same geometrical concepts apply to low-angle normal faults. Stage (a) shows the situation before displacement, with the position of the incipient fault surface marked. Thrust movement is going to occur from left to right parallel to the side of the block facing us. Looking at this face first, we can see that the fault surface runs between beds for most of its length (perhaps there is a thin, weak layer of mudrock between each of the major beds drawn), but cuts steeply upwards at two places. The steep bits are called footwall ramps, and the flat bits are called footwall flats. The fault does not reach the ground surface on this face, but presumably we would see it do so if the diagram were extended to the right. The right-hand face of the block shows the fault surface at the base of the topmost bed close to the near corner, but it turns upwards to reach the ground about a third of the way along the face. Turning our attention now to the top (ground) surface, we see that this joins a fault line running parallel to the direction of thrust movement. This sort of boundary is referred to as a side wall ramp, and along this part of its length the fault will behave as a transcurrent fault. The fault at the surface

Figure 10.3 Cross-sectional block diagrams to illustrate motion of a thrust sheet. In nature, the scale of this could be anything from a metre to tens of kilometres across. See text for discussion.

makes a right angle turn where it becomes a thrust again, then turns again to become another length of transcurrent fault.

To put it another way, the sheet of rock that is about to be transported by a thrust (we would call it a thrust-sheet) is bounded on its lower surface by a thrust fault and on its sides by transcurrent faults.

The situation after thrust movement is shown in stage (b). Looking at the near face of the block we can see that there are two places where a bed has been repeated, by being thrust over itself. Also, because the fault surface is not a plane but has steps in it, the thrust sheet has become deformed, in fact it has folded in a ductile fashion. This folding occurs only in the thrust sheet, and it does not affect the underlying units. In nature, the up-folded areas would probably become eroded away, but the changing dips of the beds exposed at the surface would demonstrate that folding had taken place, and would enable us to deduce the locations of underlying footwall ramps.

Thrust belts and collision zones

We will look at folding in more detail shortly, but first there is another important aspect of thrusts to describe. We saw in Figure 10.3 that the rocks below the thrust surface do not get folded like those above. However, very often as deformation proceeds a new thrust develops below the first. What happens is that the active thrust surface jumps forward by slicing off ramps, as illustrated in Figure 10.4.

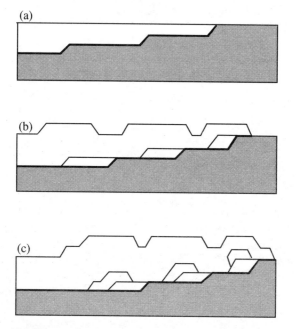

Figure 10.4 Progressive development of a thrust system, seen in cross-section. The undeformed rock below the lowest thrust is shaded. Bedding is not shown, but can be assumed to be horizontal, except where folded as a result of thrusting. The fault that is just about to become active is shown by a heavy line. Inactive lengths of thrust are shown by finer lines; note that these get folded as they are transported over footwall ramps.

In this way a series of blocks (or 'horses') bounded above and below by thrusts, called a duplex complex, develops. Rocks that were folded during motion of the original thrust sheet are liable to be folded again each time the geometry of the active thrust surface changes. Figure 10.5 shows what the final result might look like in cross-section.

Figure 10.5 could represent a 100-km-long cross-section from west to east under the eastern Canadian Rockies. These are a collisional mountain belt with thrust movement from west to east. Alternatively it could represent a 10-km-long cross section from east to west of the Moine thrust zone in north-west Scotland, where regionally metamorphosed rocks have been thrust westwards over a series of 500-million-year-old sediments.

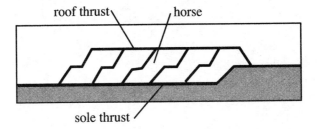

roof thrust   horse

sole thrust

Figure 10.5 Cross-section through a thrust system. This shows the sort of structural complexity that results from progressive forward propagation of the active thrusts as in Figure 10.4. A series of imbricated thrust slices, called horses, is produced, each horse being bounded by thrusts above and below. In this example, produced by thrust motion from left to right, the first horse to be formed was the leftmost one.

The biggest currently active collision zone provides an excellent example of how different kinds of fault are related. The Tibetan plateau has an average altitude of five kilometres above sea level, and has been uplifted as a result of the collision of India into central Asia. The plateau is bounded to the south by the Himalayas, which are a mountain belt that has developed by means of a series of southwards-directed thrusts. The plateau itself is crossed by many grabens (faulted valleys) delineated by normal faults running north to south, which at first sight seems odd in a region that ought to be experiencing compression directed from northwards. The explanation becomes clear when it is recognised that the plateau is both bounded and crossed by a series of transcurrent faults. Some run from south-west to north-east and are left-lateral faults, other run from north-west to south-east and are right-lateral. Tibet is buttressed by mountains on all sides except the east, and is being squeezed out eastwards by motion across these faults. The north–south normal faults are just a manifestation of Tibet's east–west extension that accompanies its north–south shortening (Figure 10.6).

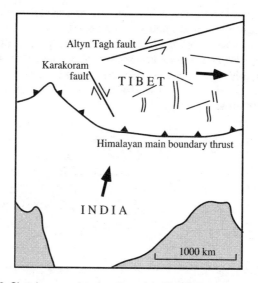

Figure 10.6 Sketch map showing how the Tibetan plateau is bounded and crossed by transcurrent faults, and north–south normal fault systems within it. Tibet is being squeezed out towards the east. The fault systems in Tibet are so large that they were not appreciated until seen from space.

So, to really understand the tectonic history of an area, the whole fault system needs to be mapped out. Sometimes it can be deduced that fault motion has reversed one or more times during a prolonged period of deformation, or that faults that seem related were in fact active at quite different times.

Brittle faults versus shear zones

The best known faults are those that were active at relatively shallow depths in the crust, a few kilometres or so, because these are the ones that are most likely to end up exposed at the surface by erosion. The rock within a few centimetres or metres of such a fault is often intensely fractured, and is described as a fault breccia. Faults therefore tend to be zones of weakness that erode away readily. Because of the fracturing associated with them, faults may act as pathways for ground water, and can be important sources of underground water in arid regions, especially where they occur in hard igneous or metamorphic rock that is otherwise impermeable.

Low-angle faults can be mapped at greater depths because they will reflect back seismic waves generated by explosions deliberately set off for the purpose. Below a depth of 10 to 20 km, it is believed that faults become ductile features. Displacement is accommodated across a shear zone, which is tens or hundreds of metres wide for moderate sized faults.

Folds

Folding is another kind of ductile deformation, and it can occur at shallow depths. We have already seen that folds are an inevitable side-effect of thrusting. The Appalachian mountains are part of a folded and thrust mountain belt that stretches from Alabama to Newfoundland and is the result of a continent–continent collision between north America and Africa that took place about 300 million years ago.

Traditionally, folds have been regarded as features produced when the crust experiences compression, however they can also form in extensional settings because of geometric constraints such as footwall ramps on low-angle extensional faults.

Figure 10.7 shows a cross-section through folded rocks, to introduce some terms commonly used to describe folds. An **anticline** is an arch-like structure, whereas a **syncline** is U-shaped in cross-section. The axis or hinge of a fold is defined as where the curvature is greatest, whereas in the limbs on either side of the axis curvature is less.

Shapes and styles of folding

Folds can take a whole variety of forms, depending on the rate of deformation, the depth, and the mechanical properties of each stratum. The strata in Figure 10.7 are folded in continuous curves, in contrast with the folds portrayed in Figure 10.3 where the changes in dip are sharp. In Figure 10.7 there is a tendency for the strata to get slightly thicker near the fold axis: this is accomplished by ductile deformation within each stratum, manifested by distortion of features within it. In extreme cases, the degree of distortion and accompanying rotation and recrystallisation can lead to the development of cleavage, especially in mudrocks, with cleavage planes being roughly parallel to the axis of the fold. To find cleavage in folded rocks should

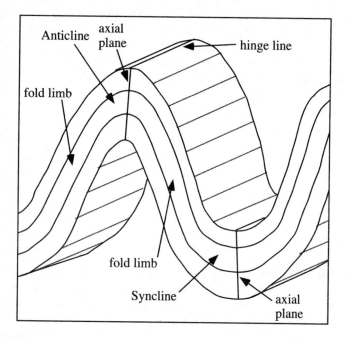

Figure 10.7 A perspective cross-section through some folded strata. This could be on any scale from centimetres to kilometres across.

come as no surprise to you, because you met cleavage as an essential element of low-grade regional metamorphism in Chapter 7, and by their very nature most regionally metamorphosed rocks have been deformed.

The axial planes of the folds in Figure 10.7 are very steep, and the hinge lines are horizontal. It is not always so. Axial planes can sometimes have very gentle dips, hinge lines can plunge downwards at any angle, and opposite limbs of a fold can diverge at a larger or smaller angle (Figure 10.8). In highly deformed regions, fold limbs are often parallel to one another, and in these cases the folds are described as isoclinal. If the axial plane dips at less than ten degrees the fold is described as recumbent (Figure 10.9). In really complex areas, folds may even be turned upside down, in which case the anticlines look like synclines, and vice versa. In such a situation the right way up has to be determined by inspection of small-scale structures such as cross-bedding.

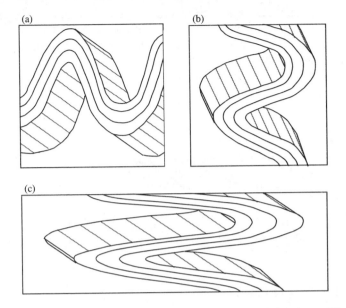

Figure 10.8 Styles of folding. (a) Upright folds, with steep axial planes. Hinge lines plunge away from the viewpoint. (b) Recumbent folds, with near horizontal axial planes. (c) Isoclinal recumbent folds, with fold limbs almost parallel.

Figure 10.9 Recumbent folds in thinly-bedded siltstones on the north Devon coast, England. These folds have a zig-zag shape, unlike the more rounded folds characteristic of limestones or sandstones.

Folds manifest themselves on the landscape in a variety of ways. It is rare for the crests of hills or mountain ranges to correspond with the axes of anticlines. It is the more resistant strata that form the high ground, as shown in Figure 10.10.

Figure 10.10 Cross-section through an eroded anticline, looking along its hinge line, showing how the strata that are the most resistant to erosion form ridges. Gentle slopes following the surface of a stratum are described as dip slopes, opposite-facing steeper slopes where a stratum has been cut through by erosion are called scarp slopes.

Unconformities

Sometimes tilted or folded strata that have experienced erosion get covered by younger deposits. These will share neither the dip nor the folding of the strata they bury. The junction between the two series of rock, which may represent a gap of millions or even hundreds of millions of years is called an **unconformity**.

An angular unconformity as in Figure 10.11 is usually easy to spot because of the pronounced difference in the amount or direction of dip between the strata below and those above. However, it can be very difficult to recognise unconformities where there is no difference in dip. These occur where the upper strata of an unfolded, untilted sequence have been eroded away in a planar fashion and then covered by younger strata, or simply where otherwise continuous deposition was interrupted by a pause during which no sediment accumulated.

Although, technically, any subsequently buried erosional surface can be called an unconformity, the term is not usually applied where the presumed time-gap is geologically short, the amount of erosion slight, and the change in environment trivial. For example, the erosional bases of the river and delta channels in Figures 9.8 and 9.10 would not normally be classed as unconformities.

Figure 10.11 Cross-sections showing younger strata deposited unconformably on the eroded anticline shown in the previous Figure. (a) No further erosion before deposition. (b) The area planed flat before deposition. In both cases the junction between the older (folded) and younger sequences is described as an angular unconformity, because of the difference in dip between the strata above and below.

One way to decide how long a time interval is represented by an unconformity is to determine the difference between the ages of the beds immediately above and below. In most situations, the most precise dating comes not from radiometric techniques such as those we met in Chapter 2, but by looking for differences in any fossils. We will examine fossils and their value in stratigraphic dating in Chapter 12, but first we will consider the Earth's physical resources.

11

PHYSICAL
RESOURCES

We rely on geological processes for all our raw materials except things that grow, and for all our energy except solar power. The geological processes of winds, waves and tides can be used to generate power, but the vast majority of power to run our industrial society comes from burning **fossil fuels**, specifically coal, oil and natural gas. Because we are using up these fuels at rates vastly in excess of the rates in which geological processes can form them, they are classed as **non-renewable resources**, along with the radioactive fuels used for nuclear power. Wind, wave and tidal power, and (to an extent) geothermal power generated from the heat escaping from shallow magma bodies, do not outstrip the rate of supply, and are referred to as renewable resources.

In addition to fossil fuels, non-renewable resources include such commodities as the **ores** from which metals are extracted, the clay used to make bricks and porcelain, the limestone used to make cement, and rock in general used to make road surfaces.

If we continue to use up raw materials at the present rate, some things will run out sooner than others. For example, there is no shortage of limestone that could be used to make cement. We might not be willing to accept the environmental impact of quarrying away beautiful limestone hills, but that is a separate issue. On the other hand, silver is being used up at a rate sufficient to exhaust the world's supply of that metal within twenty years.

The rates at which we are using up many raw materials, and the lack of recycling, are causes for concern, but there is no need to panic. The world will not actually run out of silver in twenty years. If no more silver deposits are discovered, silver will have to be extracted from deposits which are currently uneconomic to exploit. This will push the price of silver up, and encourage the use of substitutes and recycling.

Reserves and resources

This brings us to the important distinction between **resources** and **reserves**, two terms that are often misused. Their relationship is summarised in Figure 11.1. Strictly speaking, 'reserves' means only those raw materials than can be extracted profitably and legally under existing conditions. On the other hand, 'resources' should be used only in a wider sense, to describe the total amount of the raw material that could be extracted if the price were high enough, or legislation allowed, or technological advances made it feasible. Not all the resources of a particular substance have been located, and so the total resources available can usually be added to by exploration.

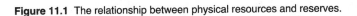

Figure 11.1 The relationship between physical resources and reserves.

The earlier statement that silver will run out in twenty years is thus rather inadequate. It should be rephrased to state that the current *reserves* of silver will be exhausted in twenty years. In practice, as these reserves are used up by extraction they will be added to by decisions to extract silver from deposits that were formerly regarded as not worth exploiting, or by the discovery of new economic deposits.

Twenty years is thus a fairly comfortable cushion, so long as new discoveries continue at a rate sufficient to keep pace with extraction. However, every non-renewable resource is liable to be exhausted eventually, so recycling and the use of substitutes will inevitably become increasingly important. In the meantime, it behoves us to extract non-renewable resources with care and sensitivity, and to consume them wisely.

In this chapter we will first look at fossil fuels, and then at other non-renewable resources such as ore minerals and bulk materials. We will concentrate on how their origins can be explained by the processes that we encountered earlier in this book.

Origins of oil and gas

Of all the fossil fuels, oil is the one that tends to make the headlines. It has become so important to the global economy that wars have been fought over it. Oil and natural gas have the same origin, and are often found in association. Moreover they are both hydrocarbons, so we will consider them together.

The raw material for hydrocarbons is dead organic matter, chiefly the remains of microorganisms and land plants. For enough of this material to collect, it is necessary to have an environment where there is a large supply of dead organic matter coupled with a lack of oxygen to stop it decaying to water and carbon dioxide. These conditions are met in those areas of the sea floor where the surface waters have a high rate of production (and mortality) of planktonic plants, which upon death sink to the bottom to be buried in fine-grained, shaley, sediment. More rarely, the floors of lakes provide a suitable environment.

Once an organic-rich shale has accumulated, it is a potential **source rock** for hydrocarbons. For this potential to be realised, burial must occur to allow heat, pressure and chemical conditions to cause diagenetic changes in the organic matter. The conversion from dead organic matter to a hydrocarbon is described as maturation.

By the time burial has reached a depth of 1 km, the temperature is likely to be about 50 °C. Under these conditions, the organic matter will be cooked into complex molecules called kerogens, which consist of carbon, hydrogen and oxygen. With deeper burial to about 6 km and temperatures up to about 180 °C the kerogens break down to chains of carbon molecules bonded with hydrogen, which is petroleum. The general formula is C_nH_{2n+2}. where C is carbon and H is hydrogen. The chains get shorter according to the temperature of maturation. Where n is 16 or more the product is a waxy solid, but shorter chains are liquid.

At slightly higher temperatures the chains get really short, producing C_4H_{10} (butane), C_3H_8 (propane), C_2H_6 (ethane), and CH_4 (methane), which collectively are known as natural gas. If the temperature gets much over 200 °C even methane breaks down, driving off the hydrogen and leaving the carbon behind as graphite. Thus hydrocarbons can form and survive only if conditions stay within a fairly narrow band. There is actually a time element too, and hydrocarbons can form and break down at lower temperatures if there is sufficient time (Figure 11.2)

Figure 11.2 Production of hydrocarbons by maturation within a suitable source-rock, showing how this depends on temperature and time.

Oil and gas reservoirs

It is one thing for hydrocarbons to form and to survive within a rock. It is quite another matter to be able to extract them. If the source rock becomes exposed at the surface, then it can simply be dug up to get at the oil within it. This is the only easy way of extracting non-liquid hydrocarbons or kerogens, but the environmental costs of this sort of quarrying are so high that most oil shales, as they are known, have not yet been exploited.

Hydrocarbons are almost always extracted by drilling. You might think that oil and gas wells are made by drilling into source rocks. This is not the case, because although shales are porous and can contain 10% or so by volume of hydrocarbons, they are not *permeable*. That is to say the pore-spaces are not sufficiently well connected to allow fluids to flow through the rock at the rate required to feed a well.

Most hydrocarbons obtained by drilling are extracted not from the original source rock but from a **reservoir rock**, into which the hydrocarbons have migrated over geological time. The criteria that must be met by a reservoir rock are that it must be both porous and permeable. Sandstones usually fit the bill, unless most of the pore-space has been filled by cement, and these are the most common reservoir rocks for hydrocarbons. A rubbly limestone, formed in a reef, may make a good reservoir rock too, especially if **porosity** and **permeability** have been increased by solution effects.

In order for hydrocarbons to be held within a reservoir rock, there must be a nearby source rock from which the hydrocarbons can have migrated, and the reservoir rock must have remained below the temperature at which the hydrocarbons break down. Once they have found their way into a potential reservoir rock, oil and gas tend to be displaced upwards by pore water, which is denser than oil. A final condition therefore, is that the hydrocarbons must be trapped within the reservoir rock in some way, otherwise they would have long since seeped out to the surface.

There are essentially two kinds of oil trap. Structural traps are when the reservoir rock is folded or next to a fault in such a way that the hydrocarbon-bearing part is bounded by impermeable rocks so the hydrocarbons cannot escape (Figure 11.3a,b). Stratigraphic traps occur below an unconformity (Figure 11.3c).

Once the appropriate part of a hydrocarbon reservoir has been drilled into, oil or gas may be driven to the surface merely by pressure. Usually about 30–40% of the oil may be recovered this way. Most of the remainder can be extracted by displacing it by pumping water or gas down a suitably positioned borehole.

Exploration for oil and gas

Oil wells are usually 1–5 km deep. It is very expensive to drill such deep holes, and so drilling is not the major technique used in *exploring*

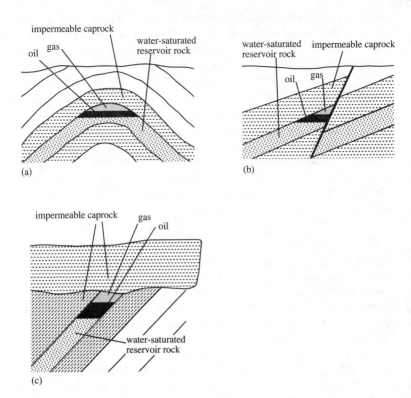

Figure 11.3 Cross-sections to show oil traps. (a) A structural trap in the form of an anticline or a dome. (b) A structural trap against a fault. (c) A stratigraphic trap, below an unconformity. In all three cases, hydrocarbons are prevented from escaping through the sides and top of the trap by the presence of an impermeable caprock. The areas labelled 'oil' and 'gas' show where the reservoir rock is saturated with these.

for hydrocarbons. Sometimes, interpretation of the surface geology may suggest that an oil trap may exist at an appropriate depth, but often these structures are too deeply buried to be apparent at the surface. In such a situation, initial exploration relies on geophysical techniques such as mapping anomalies in the local magnetic and gravity fields.

When an area appears promising, it is worth making a seismic study. This involves using explosions to generate seismic waves, which are picked up by an array of sensors. Seismic waves generated this way are not as powerful as those from a major earthquake, but they have the advantage of providing a sharp signal, generated from a known point at a known time. By interpreting the reflected seismic signals from successive rock layers, a picture of the sub-surface can be built up.

Surprisingly, seismic surveys are most easily done under water. A ship trails a kilometre-long array of sensors to record pressure variations in the water, and periodically generates a strong underwater sound pulse from a compressed air gun. Part of the energy is reflected back by the sea floor (this is the principle of echo-sounding to determine water depth), but the rest continues through the sediment, with some of it being reflected back at each interface between layers.

When an anticlinal or domal structure, or some of other kind of potential hydrocarbon trap, is revealed by a seismic survey then it is worth drilling an exploration hole. In addition to attempting to reach the hydrocarbons, much can be learned by studying the rock chippings that are deliberately flushed out of the borehole, and by lowering various electronic, magnetic and radioactive sensors down the hole to record wireline logs to give estimates of the density, porosity, composition and even the dip of bedding for each unit that is passed through. Only when a successful exploration well has been drilled and the extent and capacity of the reservoir has been ascertained is it worth sinking production wells. These may be drilled at any angle, and, especially for offshore production, a production platform may extract hydrocarbons from a three-dimensional fan of production wells that penetrate the reservoir over an extensive area.

The origin of coal

The other great fossil fuel is coal. Like the hydrocarbons, coal derives from dead organic matter, but in this case the source is mainly land plants. Several criteria must be met for coal to form. First, there must be abundant land plants living (and dying) nearby. Coal is unknown in rocks older than about 425 million years, which is when the first land plants appeared, and is particularly abundant in sequences formed at certain times between about 350 and 50 million years ago when climate, the arrangements of continents, and the evolutionary state of land plants combined to favour its formation.

The second requirement, as for hydrocarbons, is that the dead plant material should accumulate in an environment lacking oxygen ('anoxic') to prevent oxidation. Swamps meet these first two criteria very well, having abundant vegetation and a stagnant water table coinciding with the sediment surface. The organic-rich soil here is described as peat.

The next criterion for an economic coal deposit is that there should be rapid subsidence to allow a sufficient thickness to accumulate. Ten metres of peat compacts down to only one metre of coal. The final requirement is that the deposit should be buried to hundreds of metres to enable the peat to be turned into coal. All four requirements are met in delta environments, where, as we saw in Chapter 9, a swampy delta top where peat can accumulate is continuously subsiding under the weight of sediment. If the main distributary channel of the delta switches, the old delta top may be buried by marine or delta front sediments (Figure 9.10). The major coal fields of Europe, Asia and North America formed in this way, about 300 million years ago, in the latter half of the period of geologic time called the Carboniferous (sometimes distinguished by North American geologists by the name 'Pennsylvanian').

The quality or 'rank' of coal depends principally on how deeply it was buried during diagenesis. Under anoxic conditions, dead plant material is broken down by bacterial action to a woody residue called lignin. With progressive burial, volatiles such as water and methane are driven off, the deposit becomes compacted, and the amount of heat energy that can be released by burning a given weight increases. Peat has less than 50% carbon, but the highest rank of coal, anthracite, is over 90% carbon, and (weight-for-weight) contains about three times as much energy. Apart from the rank of coal, other factors that determine its usefulness include impurities within it. Clay minerals leave a solid residue (ash) after combustion, iron sulfide produces the acid gas sulfur dioxide during combustion, and sodium chloride corrodes boilers.

Mining techniques

Coal can be extracted either by deep mining or by strip mining. The latter (Figure 11.4) is economic for shallow deposits, but becomes prohibitively expensive (and more environmentally damaging) as the thickness of overburden that must be removed to get at the coal increases. Strip mining is the favoured technique in the great US coalfields of Pennsylvania.

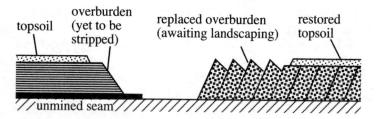

topsoil overburden (yet to be stripped) replaced overburden (awaiting landscaping) restored topsoil

//unmined seam

Figure 11.4 Area strip mining of a shallow coal seam, showing how the overburden that has to be removed to work the active face can be dumped behind, This obviates the problem of how to dispose of it, and (provided the topsoil has been kept separately) the landscape can be restored to something like its original form. The same technique has been used to mine ironstone seams in England.

In deep mining, a shaft is sunk to the level of a coal seam that is thick enough (typically 0.5–3 m) to be worth exploiting. Coal was formerly removed by men with pickaxes, but nowadays mechanical cutters are used. These make it cheaper to exploit thick seams, but have made it uneconomic to work thin coals underground. There are essentially two approaches to mining of coal, or any other sheet-like deposit (such as evaporite minerals). Each is dictated by the need to prevent the roof collapsing on the miners and equipment. The simpler and more traditional method is pillar-and-stall working, in which only about 50–85% of the deposit is removed. leaving the rest as pillars to support the roof (Figure 11.5a). It is a matter of fine judgement how thick and how closely-spaced the pillars need to be if they are to bear the weight of the roof. The alternative method is longwall mining in which an underground face is created that may be be several hundreds of meters long. A cutter operates along the seam beneath a roof supported by hydraulic jacks, and these are slid forward as the cutter passes, allowing the roof to collapse behind it (Figure 11.5b). Access is maintained by a tunnel joining each end of the face to the mine haulage roadways. As the face advances the ground surface subsides, the effects of subsidence being more extreme if the mine is shallow.

Longwall mining is the main method of extracting coal in Britain. In North America, pillar-and-stall working is favoured, although pillars are usually removed when the rest of the seam has become exhausted, allowing the roof to collapse in a controlled manner like in longwall mining.

Figure 11.5 Techniques for mining a coal seam. (a) Pillar-and-stall working. This works only if the roof is a strong rock type such as sandstone or limestone. If the roof is shale it is liable to collapse, unless the pillars are very closely spaced. (b) Longwall mining, seen looking along the face that is being worked. The cutter is moving into the page, and the roof is allowed to collapse behind it. When it reaches the end of the face, the cutter will be moved to the right, and cut back towards us.

Ores and their origins

Metals are extracted from ores. Ore is a term referring to any rock containing a sufficient concentration of a metal for it to be economic to extract. The concentration of a metal in the ore (the grade of the ore) needed to make it economic depends on the value of the metal and the cost of extracting it. For example, iron is abundant and cheap, and at present it is not worth extracting from ores containing less than 50% iron. It uses a lot of energy to isolate aluminium from its ores, so it is not usually economic to work ores less than 30% aluminium. Lead is worked from ores whose grade is as low as 5%, copper from ores around 1% grade or less, silver from ores of 0.01% (a hundred parts per million) grade, gold from ores of 1 to 10 parts per million grade, and platinum from ores as low as 0.1 parts per million grade. When several metals occur in the same ore, it can be profitable to work it for the combination of metals, even if it would have been uneconomic to work for a single metal.

Most ores are products of igneous activity. Sometimes economic concentrations are formed within a magma body when heavy crystals such as chromite (the main ore of chromium, and an important source of platinum and related metals) or nickel, iron and copper sulfides sink to the bottom. A prime example of this type of differentiated igneous ore body is the 200-km-wide Bushveld Complex to the north of Johannesburg, South Africa, which contains enormous reserves of chromium, iron, platinum, and gold. One of the most famous, and largest, nickel ore bodies is the 50-km-long Sudbury igneous complex in Ontario, Canada. Interestingly, this is thought to have originated not as the ordinary sort of crustal melt but as a melt formed when a small asteroid hit the Earth.

Other ores form not within an igneous body itself, but in the surrounding rocks. These are a result of hot watery fluids circulating through an intrusion and into the surrounding rocks as it cools. The process is one of convection, driven by the heat of the igneous body, and is described as hydrothermal circulation. Essentially what happens is that metals are dissolved out of the intrusion by the hot fluids, and deposited nearby, usually as sulfides and sometimes as oxides. The major 'porphyry copper' ore bodies in the Andes are of this type, as are the tin ores in Cornwall, England (where parts of the granite have been altered to kaolinite or china clay by the hydrolysis reactions described in Chapter 8).

A few metals, notably gold and silver, rarely form compounds and so are found as pure ('native') metal, usually as tiny flecks rather than in the fist-sized nuggets of popular myth. If the circulating fluid passes along fractures, this is where the precipitation occurs and the ores lie only within these veins rather than being disseminated throughout the neighbourhood. These fractures usually get filled mostly by undesirable minerals such as quartz, forming an ore **vein** that is mostly useless material but will be economic to mine for gold provided it contains more than about 10 parts per million.

The origin of hydrothermal ore deposits associated with igneous intrusions is summarised in Figure 11.6. Other hydrothermal ores occur at ocean spreading axes at vents such as black smokers (Chapter 5). At present it is far too expensive to mine these on the sea floor, but they have been mined in ophiolites, notably on Cyprus which was the main source of copper for the classical world.

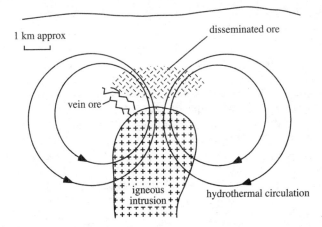

Figure 11.6 Cross-section to show the precipitation of ore bodies from circulating hydrothermal fluids. The convection system is driven by the heat of the igneous intrusion.

Not all hydrothermal ore deposits are associated with igneous intrusions. For example the lead and zinc ore bodies in the Mississippi River basin, Ireland and the Pennines of England occur within limestones, and appear to have precipitated from warm mineralised fluids expelled from deeply-buried sandstones or shales. Warm groundwaters are also suggested for the concentration of uranium into ore bodies in the Athabasca Basin of northern Saskatchewan, Canada and in the Northern Territory of Australia.

Ore exploration

Exploration for ore bodies can be done in many ways. If the ore body extends to the surface, then it may be discovered by an old-style prospector. Even when the economic part of an ore body is below the surface, hydrothermal circulation may have bleached or otherwise discoloured the rocks at the surface. Surface effects may also be discovered by analysis of visible and infra-red images recorded by satellites or airborne surveys. Other exploration techniques include mapping the magnetic field, often by an airborne survey, and looking for anomalous magnetic features that may be caused by metallic ores. When a potential ore body has been located, it must be drilled to confirm its extent and to assess the grade of the ore. Only about one in a thousand promising discoveries has all the attributes required to be turned into a successful mine.

Mining of ores

Deep mining is the standard method for most ore bodies (Figure 11.7) but hydrothermal copper deposits tend to be so extensive that they are usually exploited by surface mining. The deeper the hole grows, the more expensive this technique becomes, because the sides of the quarry must be prevented from collapsing. The only effective means to ensure this is to cut the walls back into a series of steps. Thus a progressively greater ratio of unwanted 'overburden' must be removed every time the floor of the quarry is lowered (Figure 11.8).

Figure 11.7 Cross-section through a deep mine, exploiting an inclined ore body. An adit is a horizontal or shallowly inclined tunnel that was used in this case to gain access to the upper part of the ore body, which has now been exhausted. Ore is now being extracted through deeper tunnels and raised up a vertical shaft.

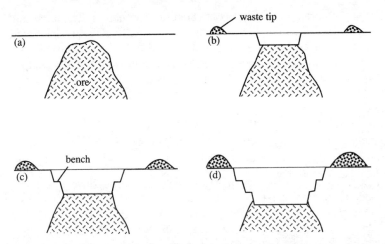

Figure 11.8 Cross-sections of 'surface mining' or quarrying of an ore body showing how the sides of the quarry are cut into a series of steps or 'benches' for stability. Each time the quarry is deepened, progressively more overburden must be removed, until it becomes uneconomic to continue. In this example, before the quarry can be deepened beyond stage (d) the waste tips must be moved.

Diagenetic and sedimentary ore bodies

Not all ores are formed as a result of hot fluids. Some can form during low-temperature diagenesis of the rock, if the chemical conditions in the water permeating the rock are correct. A good example is provided by the ironstones of Northamptonshire and Cleveland in England. These are shallow marine sandstones of Jurassic age which are cemented by iron carbonate (siderite $FeCO_3$), sufficiently easy of access and in sufficient concentration for it to have been an important ore until the latter part of the twentieth century.

Today, most of the global production of iron comes from another type of sedimentary deposit, consisting of alternating layers of silica and iron oxides and known as **banded iron formations** or **BIFs**. Virtually all BIFs are between 3800 and 1800 million years old, and are believed to have formed on the floors of shallow seas when (as we saw in Chapter 2) there was much less free oxygen around. BIF ores have been mined in the Lake Superior district of the United States since the nineteenth century. About 15% of the global production of BIF iron is from Australia, notably the Hamersley Basin in the north of Western Australia where it is mined in huge open pits.

Other ores are an end product of chemical weathering. The outstanding example of this is bauxite, which is the only globally significant ore of aluminium. This is produced under humid tropical conditions when clay materials break down under hydrolysis. For example, kaolinite, produced by the chemical reaction shown in Chapter 8, breaks down to bauxite as follows:

$$Al_2Si_2O_5(OH)_4 \quad + \quad 5H_2O = \quad 2Al(OH)_3 \quad + \quad 2H_4SiO_4$$

| kaolinite (a clay mineral) | water | bauxite ore | a form of dissolved silica |

Australia, Guinea, Jamaica, Brazil and India are the world's five major producers of bauxite ore. The only other important ores produced in this way are of nickel. This metal occurs in insoluble silicate residue in soils formed by humid tropical weathering of olivine-rich igneous rocks. About a third of the global production of nickel comes from such sources, mostly from New Caledonia, Indonesia, Cuba, Brazil and the Dominican Republic.

The remaining class of ores is **placer deposits**. These occur when heavy metal-rich grains are concentrated in sedimentary deposits. Gold is the most obvious example. All too often, in 'them there hills' where the gold is supposed to be, it occurs as tiny flecks in veins far too diluted by barren quartz to be worth mining. However, streams draining this area, and therefore transporting locally-derived sedimentary grains, can act to concentrate the gold. This is because grains of gold are considerably denser than normal sand grains and will tend to lag behind in the bed-load. When panning for gold, a prospector takes a few handfuls of sand or gravel from the stream bed and swirls it around, hoping to wash away the sand and leave the gold behind.

Most commercial gold production comes not from present-day streams, but by mining placer deposits. The world's most prolific gold-producing area is the Witwatersrand Basin in South Africa, where gold is found in pebbly and gravelly sediments that appear to have been deposited from braided streams about 2500 million years ago. The Witwatersrand conglomerates are notable also for containing placer grains of uraninite, an oxide of uranium (UO_2). Like BIFs, placer uranium ores are an indicator of the low level of atmospheric oxygen in the distant past. Placer uranium ores cannot form today, because uranium forms the compound $UO3$, which is soluble.

Non-metallic raw materials

Some raw materials are sufficiently valuable to be mined underground like ores. These include fluorspar (CaF_2) which is needed for aluminium smelting and as a flux in steel making, and various evaporite minerals such as gypsum ($CaSO_4.2H_2O$) which is used in plaster and as a filler for textiles, and potash, a general name for potassium salts (e.g. KCl and $K_2SO_4.2MgSO_4$), which is used in fertilizers and other chemicals.

Modern industrial society also requires certain materials in bulk, the most obvious being: clay and shale for bricks; limestone for cement, concrete and roadstone; sand and gravel for concrete, roadstone, and various construction purposes; and crushed hard rock in general for roadstone. These are extracted by quarrying. Sometimes a happy occurrence of mudrocks and coal seams gives two products from a single operation (Figure 11.9).

Figure 11.9 An opencast clay pit in County Durham, England. Deltaic mudrocks and silts are extracted and mixed in appropriate quantities to make bricks. Coal is a useful subsidiary product from this site; the digger is standing on the top of a coal seam, and another seam can be seen about 4 m higher.

Making good the damage

In Britain, sand and gravel come most readily from the flood plains of lowland rivers. Large quantities of these materials were deposited when these rivers were charged with a much greater than present sediment load towards the end of the last glaciation. After use, these gravel pits usually become flooded, and may be used for watersports, or turned into nature reserves.

Deeper disused quarries and mines pose more problems. Many quarries are used as 'landfill' sites, for disposing of domestic and industrial refuse. With sensitive landscaping an infilled quarry may blend perfectly into the surrounding countryside, though it annoys geologists if no rockfaces are preserved. Controls are required to limit what types of waste may be dumped in specific localities. For example, if the quarry is in a permeable limestone, then nothing should be buried there that might contaminate the groundwater. Mines are less of an eyesore, though heaps of mine waste or 'spoil' can dominate the landscape. Old mines are rarely backfilled and the most lasting legacy of mining on an area may be subsidence. Above shallow mine workings, buildings and roads on the surface may become damaged centuries after a mine was closed.

Water resources

One other great natural resource that is taken from the ground in many places is water. Water may fill the pore-spaces of permeable rocks, and can be pumped from the wells drilled into these natural underground reservoirs. If the permeable layer is capped by an impermeable layer, the groundwater may be under sufficient pressure to rise to the surface without pumping, in a so-called artesian well (Figure 11.10). When the fountains in London's Trafalgar Square were first constructed the water gushed out under artesian pressure, though nowadays they have to be pumped because extraction of groundwater has lowered the water table. Many of the oases in the Sahara desert are where artesian water carried in a sandstone layer reaches the surface.

Groundwater is essentially a renewable resource because the groundwater reservoirs are recharged by rainfall. However, in dry regions such as Arabia and the southern United States groundwater is being extracted much faster than its recharge rate.

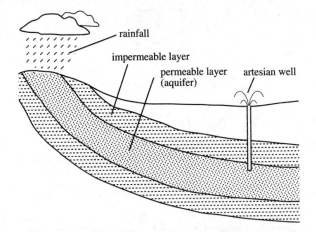

Figure 11.10 Cross-section to show how a permeable layer (known as an aquifer) soaks up water where it is exposed at the surface, This water is prevented from escaping because of an impermeable layer of rock above, but water will rise to the surface of its own accord if a well penetrates into the aquifer.

Perhaps the most priceless natural resource of all that the Earth has to offer is living organisms. Traces of life can be found in many ancient rocks. We have already discussed how coal and hydrocarbons are made from organic remains, and how limestones are usually made from the shells of living organisms. In the next chapter we will look at some of the more interesting or important organisms that can be found fossilised in rocks.

12
PAST LIFE
AND FOSSILS

For many people, **fossils** and geology are inseparable. There is no denying the thrill of discovering the remains of an organism that lived hundreds of millions of years ago. The study of fossils in their own right has taught us virtually all we know about the history of life on Earth, and has provided much of the evidence in support of the theory of evolution. Fossils are also important because traces of life in sedimentary deposits, whether as remains of the organisms themselves or in the form of burrows and the like, can be just as valuable as bedforms, grain size, and so on, in helping us determine the nature of a past environment. Moreover, whereas most characteristics of deposits in any particular environment are the same, irrespective of age, identification of varieties of fossil organisms provides the single most useful way of dating a rock sequence. This is because many species have survived for only limited periods, of the order of a few million years in the most useful cases.

The origin of life

The presence of life on Earth used to be a great mystery. However, knowing what we know now it would be a great surprise to encounter an Earth-like planet where life had *not* developed. The chemical elements required for life can be found in abundance in the oceans, and even more abundantly in hot fluids escaping from hydrothermal vents such as black smokers (Chapter 5). Experiments in the 1950s showed that amino acids (from which proteins can be built) can be made by passing electric sparks through a mixture of ammonia, methane and water. Lightning discharges into the Earth's early oceans could therefore have begun the sequence of chemical processes that ended up with life. Amino acids have also been found in carbonaceous meteorites, so an alternative possibility is that these building blocks of life were delivered by meteoritic (rocky) or cometary (icy) debris falling onto the still-cooling and sterile surface of the young Earth.

There are many steps on the road from simple amino acids to complex molecules capable of self-replication (such as DNA) and thence to cellular life. However geological time is so long and the number of molecules in the oceans so vast that there is no need to call on divine intervention to account for how life began.

The first bacteria

It seems most likely that life first got going beside underwater hydrothermal vents, fuelled by the abundant supply of chemical energy. Chemical reactions occurring across the skin of tiny bubbles could have led to the origin of the first cell membranes. There are varieties of bacteria still around today obtaining their energy chemically and able to survive only in oxygen-free conditions, which are probably little-changed from the primitive bacteria that developed beside those ancient hot vents. Today they are most noteworthy for their function in breaking down raw sewage; a rather unprestigious role for the closest relatives of Earth's first living organisms.

Once bacteria-like organisms had arisen, new varieties evolved and migrated to less chemically-rich regions where energy had to be extracted from sunlight (by photosynthesis). The simplest photosynthetic organisms today are the cyanobacteria, formerly called blue-green algae. As we saw in Chapter 2, once these organisms began to spread and multiply they began slowly to change the chemical balance of the atmosphere, in particular liberating oxygen which was at only 1% of its present level about 2000 million years ago.

The earliest direct traces of life are cells and filaments of cells, found as fossils in chemically precipitated chert and carbonate deposits. These are only about a hundredth of a millimetre across, and so can be discovered only by painstaking microscope surveys on thin, translucent, slices of rock. Most are thought to have been cyanobacteria. This cannot be proven, but it is certain they were very simple cells, lacking a nucleus and other internal structures. Radiometric dating of associated volcanic rocks shows that the oldest of these organisms must have been entombed about 3500 million years ago. As more studies are made, the record for the oldest fossil keeps being pushed back, and it would not be surprising if it ends up at about 4000 million years.

The oldest fossil structures that are easily visible date back almost as far. These are metre-scale mat-like or dome-like laminated features within calcareous sediments in the Pilbara region of Western Australia, thought to be between 3400 and 3500 million years old. These resemble modern-day stromatolites, sometimes known as algal mats, that are held together by strands of cyanobacteria. Fossil stromatolites became abundant about 2300 million years ago, and remained so for nearly 2000 million years until their decline, presumably as a result of grazing by animals and competition with more advanced plants.

More complex cells

The next step up the evolutionary ladder from bacteria and cyanobacteria came with the appearance of more advanced cells, having thick cell walls, a nucleus, chromosomes, and other internal structures. The earliest known fossil cells of this kind are about 2000 million years old, and are probably single-celled planktonic plants that lived in the sunlit upper waters of the ocean.

Multicellular organisms

Fleshy multicellular marine algae appeared about 1200 million years ago, but the first multicellular animals do not appear in the fossil record until approximately 600 million years ago. These had soft bodies and so are imperfectly preserved as fossils. A wide variety of forms arose and diversified over a short timescale (geologically speaking).

The interval of time since about 540 million years ago has long been known to geologists as the Phanerozoic Eon. (If you are unfamiliar with the divisions of geological time, you may find it helpful to refer to Appendix 3.) Phanerozoic means 'visible life', and was so named because it is not until the beginning of the Phanerozoic, during the time-period known as the Cambrian, that animals developed hard parts such as shells that do not decay upon death and which therefore fossilise easily. From then onwards, the fossil record is abundant and dominated by remains of animals with hard parts.

The time prior to the Cambrian is often informally described as Precambrian. It is divided into three Eons: the Hadean Eon from the Earth's origin until 3900 million years ago; then the Archean Eon until 2500 million years ago; and finally the Proterozoic Eon extending from the end of the Archean to the start of the Phanerozoic. 'Proterozoic' means 'earlier life', though as we have seen, we can now trace the earliest life way back through the Archean.

The major animal phyla appear

Most of the major divisions (phyla) of the animal kingdom arose during or soon after the Cambrian. These include:

1 two phyla with shells made of calcium carbonate (Figure 12.1), namely molluscs (some with two shells and some with one shell) and the less-familiar brachiopods (two-shelled organisms, with body-plans entirely different to molluscs).

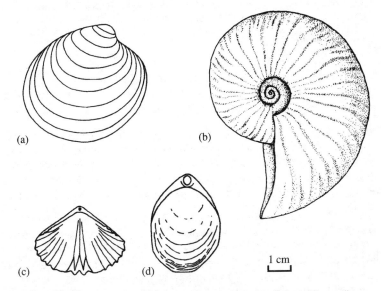

(a)

(b)

(c)

(d)

1 cm

Figure 12.1 Bivalve (a) and ammonoid (b), two distinct varieties of the mollusc phylum, and (c), (d) two brachiopods, which although at first sight similar to bivalve molluscs belong to a different phylum.

2 a phylum known as the arthropods, having jointed chitinous external skeletons, most famously represented in the fossil record by trilobites (Figure 12.2), whose present-day distant relatives include crustaceans (e.g. crabs and lobsters), spiders and insects.

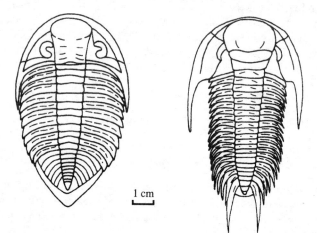

1 cm

Figure 12.2 Two trilobites, representatives of a group of arthropods that first occurred in the Cambrian and became extinct during the Permian. These two examples show the complete articulated external skeleton, except that the fragile antennae and legs are missing. Compound (insect-like) eyes can be seen on either side of their heads.

3 a phylum known as the echinoderms, characterised by a multi-element skeleton of calcium carbonate, often with five-fold radial symmetry, that includes sea urchins, star fish and crinoids.

4 the hemichordate phylum, obscure and unimportant today, but represented in the Paleozoic by a group of floating colonial organisms known as graptolites (Figure 12.3) whose rapid evolution and widespread distribution in marine sediments make them exceptionally useful for dating rocks, especially during the Ordovician and Silurian Periods.

Another important phylum, the cnidarians (or coelenterata), which includes the jelly fish, appeared in the late Precambrian. These are very rare in the fossil record until about 460 million years ago when corals first appeared. The reason that corals are well represented in the fossil record is that the soft coral polyp encloses itself in a hard

Figure 12.3 Two graptolites; the one- and two-branched forms on the left and right are characteristics of species abundant during the Silurian and Ordovician respectively. Only the hard parts are shown here, consisting of a series of cups arranged along one or more branches; before fossilization these were made of a protein called collagen. In life, a separate animal lived in each cup, catching food particles as the colony drifted though the water.

calcium carbonate cup. This is readily preserved after death, especially in the case of massive colonial corals where adjacent cups protect each other.

The chordate phylum, which includes the vertebrates to which humans belong, was around from the early Cambrian too, although initially it is insignificant in the fossil record. However, a flourishing fauna of jawless fish developed about 500 million years ago, and bony fish with jaws followed about 80 million years later.

Vascular plants and life on land

About 420 million years ago plants containing tubes for transport of water and nutrients (vascular plants) began to spread onto the land, and were shortly followed by insects and other arthropods. Some fish developed the capability to breathe out of water, and about 370 million years ago some of their descendants evolved into the first primitive amphibians, followed within 20 or 30 million years by the earliest reptiles.

Animal life on land was dominated for a long time by reptiles, notably the dinosaurs that flourished throughout the Mesozoic Era (encompassing the Triassic, Jurassic and Cretaceous Periods). At the same time, their swimming reptiles such as ichthyosaurs (Figure 12.4) and plesiosaurs were among the largest marine creatures. Other relatives of the dinosaurs, the pterosaurs, took to the air.

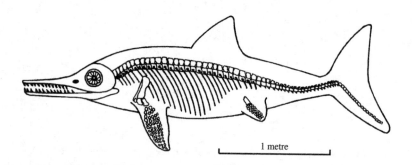

Figure 12.4 An ichthyosaur, a marine reptile from the Mesozoic Era. Usually only the bones survive, but sometimes an impression of the skin is preserved, which has enabled the dorsal fin and the upper part of the tail to be shown in the outline drawn around the skeleton.

Mammals were present on land for most of the Mesozoic, though they were small and insignificant creatures, known mostly from their teeth which are the only parts of such small-boned creatures to fossilize well. The earliest known bird, *Archaeopteryx*, is of late Jurassic age (about 170 million years old). It is known from fossils found in the fine-grained Solnhöffen Limestone of Germany, and would probably have been dismissed as a winged reptile but for the impressions of its plumage within the rock.

The first flowering plants did not evolve until about 100 million years ago, midway through the Cretaceous, and did not surpass the conifers in importance for a further 30 million years or so. Back in Carboniferous times (360–290 million years ago), the trees responsible for forming most of the world's coal deposits belonged to neither of these groups, instead vegetation was dominated by giant relatives of the club mosses (lycopods), which today are small and inconspicuous.

The mammals take over

It was not until the dinosaurs disappeared at the end of the Cretaceous period (65 million years ago) that mammals had their chance to diversify, and become the dominant group of land animals. They also spread into the sea, with the ancestors of modern whales first appearing about 50 million years ago.

The fossil record of human ancestors is not good. Complete skeletons are unknown, and most of the story has to be pieced together from skulls (usually incomplete) and other bone fragments. The first human-like apes such as *Australopithecus*, appeared about 5 million years ago. The first bones recognised to be from species belonging to our own genus, *Homo*, are about 2.4 million years old and *Homo sapiens* does not appear until about a quarter of a million years ago.

This timescale should make it clear to you that movies and cartoons showing humans hunting dinosaurs (or *vice versa*) are very wide of the mark indeed, because they overlook the matter of a 60-million-year gap! On the other hand, that other well-known extinct animal, the mammoth (essentially a hairy elephant) was very much a human contemporary. Mammoths feature prominently in cave paintings, and the last ones died out as recently as 12 000 years ago. It is debatable to what extent their final demise was due to hunting or to the loss of their habitat because of climate change.

Mass extinctions and adaptive radiations

But why did the dinosaurs become extinct at the end of the Cretaceous, after having been so successful for so long? This is a question that has taxed geologists for many years. Various explanations have been proposed, including climate change, volcanic eruptions, and mutations caused by cosmic rays. Probably several factors were at play, and the dinosaurs appear to have been in decline for at least a few hundred thousand years before the knockout blow was delivered by the impact of a large meteorite. The probable impact site is the 180-km-diameter Chicxulub crater off the Yucatan coast of Mexico, which was discovered by geophysical techniques in the 1980s.

Such an impact would have injected a vast amount of dust into the atmosphere. Supplemented by smoke from widespread burning, this would have reduced the sunlight reaching the surface to such an extent that photosynthesis became impossible. Plants would have died, and the whole food web depending on them would have fallen apart. Recovery of the vegetation on land and marine planktonic plants would have begun within a few years (from seeds and spores), but many groups of animals and some plants became extinct. The effects were much wider than merely the demise of the dinosaurs. Many important marine groups vanished, notably the ammonoids

(Figure 12.1b), reef-building molluscs called rudists, and the large marine reptiles. Most species of marine planktonic animals of the kind known as foraminiferans disappeared too, although the group as a whole recovered.

Because it was global in extent, and has effects that can be seen among organisms living in virtually all environments, this end-Cretaceous episode is described as a 'mass extinction'. The fossil record shows several mass extinctions during the Cambrian Period, and a more dramatic one at the close of the Ordovician Period. Because there was no life on land yet, these mass extinctions are purely marine. However, a mass extinction at the end of the Permian Period (and thereby marking the end of the Palaeozoic Era) wiped out whole groups of land and marine animals, and seems to have been an even more severe event than the end-Cretaceous mass extinction.

The causes of these earlier mass extinctions are uncertain. Statistically it is unlikely that the end-Cretaceous meteorite impact was the only one of that size to have happened in the Phanerozoic, so that it is likely that some of the earlier mass extinctions were triggered in the same way.

What is certain is that after each mass extinction, those varieties or organisms that had survived were able to take advantage of the demise of their competitors to adapt, through evolutionary change, to new lifestyles. This is described as adaptive radiation.

The sudden blossoming of the mammals at the start of the Cenozoic Era is a good example. The disappearance of the dinosaurs meant that on land there were no longer any large herbivores or predators. The sudden lack of competition made it possible for the tiny mammals to diversify, and within a few thousand generations some groups that had moved into new environments were well on the way to evolving into different species better suited to take advantage of the new opportunities. A classic case of the meek inheriting the Earth, if ever there was one!

How fossils form

A fossil is any sort of remains of a past organism. Footprints and burrows count as fossils, and these are termed trace fossils. On the other hand, an impression of the body where it came to rest after death, or some part of the body that has actually been preserved, is described as a body fossil. It is rare for fleshy material to survive – it is usually broken down by bacterial decay or eaten by scavengers before there is time for it to be buried, which is why the vast majority of fossils are of the bony, shelly, or other hard parts of an organism.

To follow the stages by which an organism can be turned into a fossil, we will take the example of a bivalve mollusc, like a cockle or some similar two-shelled shellfish familiar to you from visits to the seaside. To become a fossil is a long and involved process, and is an unlikely outcome for most individuals.

The most obvious requirement for an organism to end up as a fossil is that it must become buried. There are many mischances that can prevent this happening. If our mollusc has the misfortune to be killed by a predator, then the soft parts will be eaten. The predator will probably have to break at least one of the shells to gain access to the flesh. so it is unlikely that both shells will survive intact. Even if they do, the two shells may become separated. On the other hand, if the organism dies a natural death the soft parts of the body will almost certainly rot away but neither shell is likely to be damaged. The shells are held together at the hinge by fibrous tissues, which are among the last to rot so there is a reasonable chance that the two shells will remain held together until they become buried.

A lot depends on where the creature is when it dies. If it is the sort of bivalve that lives in rivers, then after death the empty shells are likely to be rolled downstream and will pretty soon get broken up. If it is a marine bivalve that lived and died on a rocky shore it will probably get smashed in the next major storm. If it dies on a beach, it would get rolled around by the waves, the two shells would probably separate, and become broken and abraded, ending up as unrecognisable fragments of shell of the sort that make up some kinds of limestone. *Only* if the organism dies in a low energy environment, somewhere quiet enough for muddy sediment to be accumulating, does it stand a high probability of escaping mechanical breakdown of the sort described above, long enough to become buried by sediment.

However, there is a very important exception to the previous argument, which is that many species of bivalve (including cockles) spend most of their time burrowing within the sediment. If one of these dies in its burrow, then it is already buried, and the two shells will be held together by the weight of the surrounding sediment even after all the connecting tissues have rotted away. An instant fossil, you might think, but this is far from the end of the process (Figure 12.5). Very few fossils, and only relatively recent ones, occur as unchanged shell material.

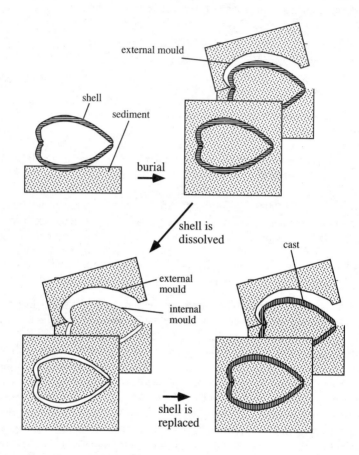

Figure 12.5 Stages in the fossilization of the hard parts of an organism. A bivalve mollusc is illustrated (shown in cross-section), but the same processes can affect any kind of shell, tooth or bone.

The first thing to happen to a buried bivalve, minus all its soft parts, is usually that some sediment will find its way inside the shells. Once the sediment has hardened (by the diagenetic processes described in

Chapter 9), the rock contains an impression of the outside and the inside surfaces of the shells, known as an external mould and an internal mould respectively. Even if, as often happens, the shells are subsequently dissolved away, these moulds remain, and many fossils are of this type. If the shell survives, usually it recrystallises in some way, so that its original microscopic crystalline structure is lost. It often happens that the whole shell is dissolved by ground water, but if conditions change then the shell-shaped void in the rock can be re-filled, to produce a cast whose mimicry of the surface texture of the vanished shell is limited only by the quality of the impression that it left in the sediment. It can be very difficult to distinguish a cast from a recrystallised shell if these are of the same composition (calcium carbonate, in the case of molluscs).

The steps described above apply to most types of fossil, with various changes of emphasis. An ammonoid has only a single shell and in that respect the complete hard parts stand more chance of being preserved together than in the case of a bivalve mollusc, but on the other hand ammonoids did not live in burrows, so in that respect the chances of becoming fossilized are reduced. Sea urchins and other echinoderms have a skeleton consisting of a large number of fragile plates, which are prone to separate after death, so low energy conditions and rapid burial are more important than for bivalve molluscs (fortunately many kinds of sea urchin were burrowers, so these can be found beautifully preserved). Trilobites and other arthropods have segmented skeletons, so rapid burial is especially important for these, too, for the hard parts to remain intact, moreover the chitin of which the skeleton is made does not survive over geological time, and must be replaced by some form of mineral if anything other than a mould is to be preserved. In vertebrates, the teeth are the most robust elements, so these are the bits that are most commonly found as fossils. The bones are less commonly preserved, and when they are, the original porous calcium phosphate structure has usually been invaded and replaced by new minerals that have precipitated from groundwaters.

Exceptional preservation

The foregoing does not mean that flesh cannot be preserved, but it is the exception rather than the rule. Examples include complete 20-thousand-year-old mammoths entombed in ice in Siberia, and 100-million-year-old-fish in northern Brazil, which were killed by highly saline waters that caused phosphate minerals to be precipitated inside their bodies within a few hours of death, so that the fine-scale cell structure is sufficiently well-preserved to warrant study using high-powered electron microscopes.

Perhaps the best known circumstances for a whole organism to be preserved is when it has the misfortune to become trapped in tree resin, which then hardens to form amber. Delicate insects and even small frogs can be found within amber. It is even true that the blood cells from a blood-sucking insect's last meal can end up being preserved in this way, though cloning a dinosaur from the DNA within these cells, as in the film *Jurassic Park*, remains a very distant prospect.

Unusual preservation may also occur when environmental conditions change so suddenly that a whole living community is wiped out. A classic example of this is found in the middle-Cambrian Burgess Shale of British Columbia, where animals living on a muddy sea floor were now and then transported by submarine landslides into deeper water, lacking in oxygen, where they died and were immediately buried. The lack of oxygen kept scavengers at bay and inhibited decay, with the result that entirely soft-bodied animals are preserved, as flattened organic residues. This sort of 'mass mortality' deposit is important because it gives us a snapshot of a living community. In the case of the Burgess Shale, only about one fossil in twenty is a shelly organism, which is a salutary reminder of how heavily most of the fossil record is biased in favour of robust organisms with easily-preserved hard parts.

Plant fossils

Plants have no shells or bones, so you might expect them to make rather poor fossils. However, unlike animals, land-plants have rigid cell walls made of cellulose. As a result, woody tissues and even leaves retain their form after death and burial long enough for the spaces inside the cells to be filled by minerals, such as chert, that precipitate from groundwater. This is how petrified wood is formed.

An alternative mode of preservation is when plant material is buried in water-logged soil or mud with a low oxygen content. In this case the leafy or woody material becomes converted to carbon or a carbon-rich compound.

A major problem in understanding fossil plants is that it is very uncommon to find a whole plant preserved. Roots (as you might expect) are relatively common, but these are rarely found still in association with leaves or cones. This has led to different parts of the same plant being given different fossil names!

On the microscopic scale, pollen and spores are often well preserved, because their coats are made of sporopollenin, a substance that is remarkably resistant to decay. In sediments up to a few million years old, when many plant species were the same as those today, or their very close relatives, analysis of pollen is a very important way of establishing the environmental conditions. For example, changing tree communities in northern Europe towards the end of the last glaciation are well documented by the pollen record, which indicates climatic fluctuations between warm and cold superimposed on the general warming trend.

The use of fossils in stratigraphy

It is unusual for sediments to contain any materials that have grown within them and whose age can be dated by the radiometric dating techniques we met towards the end of Chapter 2. If a sediment is interbedded with a series of lava flows or layers of volcanic ash, then radiometric dating of these volcanic rocks can be used to bracket the age of the sediment in between. However, convenient layers of igneous rock are often absent from sedimentary sequences, and anyway radiometric dating requires sophisticated laboratory facilities that a geologist is unable to call on while in the field. This is where being able to identify fossils pays off, because if you already know the time span over which a particular kind of fossil organism was alive (because somebody else has been able to determine this in a location where fossil-bearing sediments are interbedded with radiometrically datable volcanic rocks), then you can tell the age of a rock merely by recognising the fossils within it.

To take a simple example, trilobites are pretty distinctive (Figure 12.2), and there is not much risk of mistaking one for something else. You have already seen that trilobites first appeared during the Cambrian and became extinct during the Permian. This means that if you find a trilobite in a rock then its age must lie within these limits. This is a period of about 300 million years, which is of course rather a long time. However, during this period many different species of trilobite evolved and then died out, with an average species lifetime of around 10 million years. The two trilobites shown in Figure 12.2 are clearly different from each other and although they are both trilobites, these two are not very closely related. The differences between closely-related species are usually more subtle, and it requires well preserved specimens and an expert's eye to make an unequivocal diagnosis enabling the age of the sediment to be pinned down to within ten million years.

It is possible to improve on this if, in the same layer of sediment, you are fortunate to find fossils of two species that have overlapping age ranges. This can be seen with reference to Figure 12.6. If you found only species C in a bed of rock it would not pin the age down very closely, because this species was alive from 500 to 467 million years ago. Species B on its own would be more diagnostic, because it was around for a shorter period (493–478 million years ago), and species D on its own would give an even more precise age (470–460 million years). However if you found species B in the same bed as species A, then the age of this bed could be pinned down to the two-million-year period (480–478 million years ago) where the age ranges of these two species overlap. An almost equally precise date could be inferred if species C and D were found in the same bed.

Figure 12.6 Diagram showing the age ranges of four species of fossils (heavy lines). See text for discussion.

The above shows how we can infer quite precise ages of sediments, especially if we can use several fossils in combination. These principles can be extended to match up the ages of sediments deposited in areas that are remote from each other. For example, if species A and B were found together in North America and in Europe we would know that these two unrelated deposits were of almost exactly the same age (this would be true even if we did not know the absolute ages from radiometric studies). Correlating between rock sequences in this way is called biostratigraphy.

Some fossils are better than others for biostratigraphic correlation. The best ones are called zone fossils, or index fossils. A good zone fossil:

- has easily recognisable features that enable it to be distinguished from other species
- underwent rapid and irreversible evolutionary change
- is numerically abundant
- lived in a wide range of environments
- lived across a wide geographical area.

Trilobites satisfy the first three points, but usually fail on the last two. Most trilobite species had fairly specialised lifestyles, and not being good swimmers their geographic ranges were limited (for example, entirely different species occurred at the same time on opposite sides of major oceans). The best zone fossils tend to be organisms that swam or floated passively in the oceans and whose hard parts, upon death, sank to the bottom and are therefore preserved in both deep- and shallow-water marine sediments. Prime examples are graptolites (especially useful in the Ordovician and Silurian), ammonoids (especially in the Jurassic and Cretaceous), and the microscopic planktonic foraminiferans (with calcium carbonate shells), notably during the Cenozoic.

There is one major pitfall to beware when using fossils for dating, which is that fossils can be eroded from an old sediment, transported somewhere else, and then reburied. 'Derived fossils' such as these will indicate erroneously old ages. Usually, however, derived fossils can be recognised as such because they will show signs of abrasion and other damage during transport. One way to be sure that a fossil belongs in the rock where it is found is if it is in 'life position', as is the case, for example, with a bivalve mollusc that died in its burrow or a coral that is still attached to the rest of the reef.

Now that we have completed our overview of life on Earth, and seen how fossils can be used to correlate ages of rocks between different parts of the globe, we can turn our attention, in the next chapter, to global Earth history.

13

A SHORT HISTORY
OF THE EARTH

At the end of Chapter 2 we followed the story of the Earth's origin as far as its birth as a hot, young planet. We also saw how the atmosphere has changed over time. Now we will trace the other main events in the history of the Earth, and will risk a few predictions about the future.

We will resume our story by imagining the young, hot Earth, about 4500 million years ago. Most of the outer part is likely to be molten, because of the heat liberated by major impact events. The surface skin has cooled and solidified, but it is punctured every so often by upheavals when primordial gases escape from the interior, and by meteorite impacts from above. The magma ocean (as it is called) below the skin is convecting vigorously, and from time to time the convection currents rip the solid skin asunder, and a vast tract of cold, and therefore dense, skin founders beneath a fresh cover of magma. A terrible place, but eventually it will turn into our home.

The Earth's first crust

The solid skin of the infant Earth is not what most geologists would recognise as a planetary crust, because it is merely the chilled equivalent of the magma beneath it, having the same composition. However, during the next 100 million years or so, the magma ocean would have cooled down sufficiently to allow crystals to form within it. The first would probably be a calcium-rich variety of feldspar known as anorthite. These would be less dense than the remaining magma, and so would tend to rise upwards, ultimately displacing the chilled skin to form the Earth's first true crust.

The remains of the underlying magma ocean would now have a different composition, because of the loss of those elements that were concentrated into the anorthite crystals. We can now refer to this as the mantle. Deeper still, the process of core-formation (see Chapter 2) was probably essentially complete by now.

None of the anorthite-rich 'primary crust' survives on the Earth today, but its equivalent can still be seen in the pale highland regions of the Moon. These are formed of a rock type called anorthosite, which is dominated by anorthite crystals.

Throughout this crust-forming process, the surface would have suffered from a continuing rain of impacts by chunks of debris. Although there were no planetary embryos left, it is clear from looking at craters on the lunar highlands that there were still plenty of small- to medium-sized planetisimals, capable of forming craters up to hundreds of kilometres across. There were even a few larger plantesimals whose impacts created 1000-km-wide basins on the Moon as recently as 3900 million years ago. The size of these largest impactors can be deduced by bearing in mind that collisions within the solar system usually take place at speeds of a few tens of kilometres per second (not per hour!), and that a crater produced by such an impact is about thirty times the diameter of the impacting body.

Dating of lunar craters also shows us that the rate of impact events declined sharply after this time, and had reached something like its present level by about 3800 million years ago. This marks the exhaustion of the widely-dispersed debris left over from the birth of the planets. Subsequent impacts have been mostly by stray comets or by rocky or iron-rich chunks flung out of the asteroid belt by chance events. We know some of these hit the Earth as well as the Moon, because there are a hundred or so well-documented impact craters on the Earth. However, the vast majority of the Earth's impact craters have been removed by erosion, and tectonic and volcanic activity.

To return to the Earth well before 4000 million years ago, with its newly-formed primary crust of anorthosite and still in the throes of the early bombardment, it is likely that the mantle was convecting vigorously, even though by now it was mostly solidified. Above sites where convection cells diverged, the primary crust would be rifted apart, and partial melting of the upwelling mantle would yield basaltic crust similar to today's oceanic crust. As we saw in Chapter 4, oceanic crust is recycled at subduction zones. No Archean oceanic crust has survived, except for some debatable and deformed fragments caught up in collision zones. However, its presence can be inferred because partial melting of oceanic crust at subduction zones appears to have led to the growth of continental crust. Andesitic and other magmas above sites of subduction impregnated and overprinted the primary anorthosite crust, to the extent that no recognisable remnants of this have survived (Figure 13.1).

Figure 13.1 Hypothetical cross-sections through the crust about 4200 million years ago showing how anorthosite primary crust could have been replaced by continental crust. In (a), andesitic magmas produced by partial melting of oceanic crust are beginning to invade the primary crust, and by stage (c) the process is complete.

The oldest known continental crust consists of metamorphic rocks about 3800 million years old, with the best documented examples being near Isua in south-west Greenland. The oldest unit here is the

Amitsoq gneiss (apparently a metamorphosed granite), which is over-lain by slightly younger metamorphosed sediments (including tiny flecks of carbon that appear to be biological in origin) and volcanic rocks. Examples of similar age occur in northern Michigan, in South Africa (the Sand River gneiss), and Western Australia. The oldest rocks in Europe are 3500-million-year-old gneisses at Gruinard Bay in northwest Scotland. Metamorphic rocks must have had precursors, so it is certain that the earliest continental crust must have formed even earlier. Recent advances in analytical techniques have enabled individual mineral grains to be dated radiometrically. The age record is held by 4200-million-year-old grains of zircon (a very hard and resistant mineral) collected from within early Archean metamor-phosed sediments (in fact a quartzite) from Mount Narryer in Western Australia. Although the sediments in which these grains ended up are 'only' 3600 million years old, the grains themselves evi-dently began life by crystallising in an igneous rock some 600 million years before. Whether this rock was true continental crust or primary anorthosite crust will never be known.

With the appearance of the first rocks, we need place less reliance on speculation and analogy with other planetary bodies. Subsequent developments in Earth history can be read directly, although the record is patchy at first. During the early Archean, the picture is one of relatively small 'continental' masses, only about 50 km across. being carried around by the rapid opening and closing of intervening areas of oceanic crust. This oceanic crust differed from today's because the mantle that it was extracted from was about 200°C hotter than now. As a result, rather then being basaltic in composition, it was poorer in silica and richer in magnesium, and was of a rock type known as komatiite.

The edges of the small continental masses were marked by volcanism, presumably because of subduction-related melt generation where they over-rode the oceanic crust. Erosion in the interiors shed sedi-ment into the oceans, and each time two continental masses collided the sediments and volcanic rocks caught up between were deformed, and deeper parts of the crust were metamorphosed.

Archean plate tectonics?

It would not be wise to regard the Archean opening and closing of oceans and the associated movement of continental masses as plate tectonics in the sense we know it today. Both continental and oceanic crust, and indeed the lithosphere as a whole, were probably thinner than now, and we do not know whether these components behaved as rigid plates as they do today.

The growth of cratons

Throughout the Archean the volume of continental crust grew by igneous activity wherever oceanic crust was consumed. Between about 3000 and 2500 million years ago, many of the hithero individual tracts of continental crust became welded together by collisions. Many of these amalgamations of crustal elements developed into regions that have remained stable ever since. These are described as **cratons** or **shields**. Each craton has two elements: the original tracts of continental rocks, which are loosely described as 'granite', and the deformed belts of volcanic and sedimentary rocks, collectively termed 'greenstone belts', that were caught up in the collision zone between each 'granite' block. Such an assemblage of crustal material is described as a granite-greenstone terrain (Figure 13.2).

Figure 13.2 Sketch map of part of a granite-greenstone terrain in Western Australia, near Marble Bar. This is part of the Pilbara Block, a 500 km wide craton that extends inland from Dampier and Port Hedland. 'Granites' are indicated by a cross ornament, and the trend of deformation structures in the intervening greenstone belts is shown by fine lines. The two dykes of Proterozoic age, and which post-date the assembly of this craton, are shown by thick lines.

The largest granite-greenstone craton is the Canadian or Laurentian Shield, which makes up most of the eastern half of Canada from the Arctic Ocean to the Great Lakes (extending south of the border into the Minnesota–Michigan region). Most of the rest of North America is younger rocks that became accreted around the shield as a result of subsequent deposition of sediments, volcanic activity and plate collisions. Each of today's major continents has a cratonic element within it. Australia has the Pilbara craton (Figure 13.2) and the Yilgarn craton to its south, Africa has the Kaapvaal and Zimbabwe cratons in the south and several others further north and west, Europe has the Baltic shield (Finland and adjacent parts of Russia), Asia's largest is the Aldan shield in the far south-east of Russia, and South American cratons occur in Brazil and Venezuela.

The Proterozoic Eon

Our story has now reached approximately 2500 million years ago, which marks the start of the Proterozoic Eon. Various authorities estimate that by this stage the volume of the continental crust had reached somewhere between 40 and 80% of its present value. However, Archean cratons do not occupy anything like this proportion of today's continental surface. This is because the cratons represent only those regions of Archean crust that have remained stable ever since, and which happen not to be buried by younger sediments.

We will not attempt to survey Proterozoic history in detail. It is both complex and poorly-understood, though it becomes easier to unravel as we approach the Phanerozoic (beginning 540 million years ago). Essentially, Proterozoic history is one of rifting apart and drifting together of cratonic blocks into new configurations, and gradual increase in the total volume of continental crust by means of volcanic and intrusive activity.

By now, so far as can be told from the few deformed slivers that remain, the oceanic crust had become basaltic rather than komatiite in composition, as a consequence of the mantle having become less hot. Some rifts never quite managed to split apart, and become sites of deposition for large volumes of sediment. Most these sediments are of types familiar today, except for the lack (or scarcity) of fossils. However, because the atmosphere was still poor in oxygen some sediments were kinds that do not form today. Notable among these are the banded iron formations (BIFs) that we referred to in Chapter 9. In fact, the commercially important Hamersley basin, where about 15% of the world's iron is produced, overlies the Pilbara craton south of the area shown in Figure 13.2.

Phanerozoic plate configurations

We will now wind the clock on to Cambrian times, beginning 540 million years ago at the start of the Phanerozoic Eon. Experts agree that well over 90% of the Earth's continental crust had been created by this time. From now on, the record of the rocks offers enough clues for us to fit together a fairly complete story of events, although there is not space to trace it fully here.

It is clear that in Cambrian times the distribution of continental crust was radically different from today. Evidence for this includes study of paleomagnetism, which is magnetism trapped within rocks (especially volcanic rocks) at the time of their formation. When a rock cools down, it picks up an internal magnetisation that is parallel to the Earth's magnetic field at the time. This enables us to determine both its orientation relative to the pole and its latitude at the time of its formation. It does not tell us whether it was in the northern or southern hemisphere, because, as we saw in Chapter 4, the polarity of the Earth's magnetic field has flipped many times in the past. However, this ambiguity can usually be resolved by back-tracking from the present location, provided that a region has suitably magnetised rocks from several intervals of geologic time.

Other evidence comes from the similarity of rock types in continents that are now widely separated. Sediments are particularly useful in this respect, but intrusive rocks and deformed belts can also be informative.

Paleomagnetism may tell us that two continents were at the same latitude at a certain time, and if they do not share any geological features of this age we can conclude that they were not joined. However, this information does not tell us how far apart the two continents were. Their east–west separation is unconstrained because paleomagnetism gives no information on longitude. What makes the Phanerozoic so special is that there are abundant fossils. If fossils are similar in our two continents, then we can conclude that they were fairly close together, whereas if the fossils in the two regions are unrelated species it is likely that they were separated by a wide expanse of ocean.

To illustrate Phanerozoic Earth history, we will concentrate on the changing relationships between North America and northern and western Europe. In Cambrian times, most of the present continent of North America had been assembled. The bits still missing were various exotic terranes forming the present western coast (see Chapter 4), most of the present east coast from Newfoundland to the Carolinas, and also Florida. On the other hand, the continent was augmented by the presence of Greenland (firmly attached to what is now northern Canada), and by what is now Scotland and the north-western half of Ireland, which were somewhere close to Greenland. Clearly this continent was not identical to North America, and so it is generally given the name Laurentia, referring to the Laurentian shield at its heart.

Paleomagnetic studies show that Laurentia lay astride the Equator during the Cambrian, and that the part that would later become North America was rotated about 90° clockwise relative to its present orientation (Figure 13.3).

Figure 13.3 Sketch map of the whole globe in the middle Cambrian, about 520 million years ago, showing the distributions of the continents. Present day coastlines are drawn to aid identification, but they do correspond to shorelines at this time. Abbreviations: Af, Africa; Ant, Antarctica; Ar, Arabia; Aus, Australia; Fl, Florida; Ind, India; Kaz, Kazakhstania; NZ, New Zealand; SAm, South America. Many tracts of continental crust consisted of more than the names shown here imply; for example China includes much of south-east Asia, and Iran includes Turkey. The named regions within Gondwana had no separate identities at this time, and are outlined merely to aid recognition. England was a component of Avalonia, which consisted of several microcontinents, most of which are now exotic terrains on the east coast of North America. (Modified from various sources)

The nearest other continent at this time consisted of what is today Scandinavia and most of Russia west of the Urals. This continent was built around the Baltic shield, and is referred to as Baltica. Paleomagnetism shows that Baltica was roughly half way between the Equator and one of the poles, and it is most likely that it lay in the southern hemisphere.

To complete the Cambrian global picture, Africa, Florida, South America, Antarctica, Australia, India, Tibet, Arabia, 'Iran', and south-ern Europe were joined in a 'supercontinent' known as Gondwana, which straddled the Equator but appears to have been well out of the way of Laurentia. There were three smaller isolated continents con-sisting of what are now Siberia, the Kazakhstan region of central Asia, and China plus south-east Asia. Between Gondwana and Laurentia lay several microcontinental fragments, collectively known as Avalonia, consisting of what would later become the Carolinas, Newfoundland, Nova Scotia, the British Isles (except north-west Ireland and Scotland), and parts of northern France and Belgium.

Throughout the Cambrian, the geology of Laurentia and Baltica have little in common. Fossils in marine sediments on the facing margins of the two continents are generally unrelated, whereas the unity of Scotland with North America is confirmed by the presence of identical species of trilobites and other animals. During the succeeding geologi-cal periods, the Ordovician and Silurian, the fossils found in Laurentia and Baltica gradually become more similar. Clearly the ocean between the two continents was becoming narrower, and the two continents had collided by the early Devonian (Figure 13.4). This ocean is now vanished, but it is given the name Iapetus.

Various bits of Avalonia accreted onto Laurentia from the middle of the Ordovician (about 470 million years ago) onwards. English fossils became more similar to Scottish ones during the Ordovician and Silurian, showing that the English terrane was sliding into the Laurentia–Baltica collision zone. Its speed looks prodigious if you compare Figures 13.3 and 13.4, but it works out at about 10 cm per year which is comparable with present-day speeds of relative plate movements.

The geology of the Avalonian terranes contains many indicators of their histories. For example, many of them contain andesitic volcanic rocks of the kind generated above subduction zones, showing their essentially island arc character prior to accretion onto Laurentia. The Ordovician volcanic rocks of the English Lake District are a good

Figure 13.4 Map in the style of Figure 13.3 showing the continents in the early Devonian, about 400 million years ago. See caption to Figure 13.3 for explanation. (Modified from various sources)

example of this, and the deformed stack of sediment that accumulated in the related subduction zone trench can be seen today in the Southern Uplands of Scotland. The edge of Laurentia that bore the brunt of the collisions was deformed by folding and thrusting, and a number of slices of the Iapetus ocean floor survive as ophiolites thrust onto the continent, notably several in Newfoundland.

By the end of the Devonian, Laurentia, Baltica and all the bits of Avalonia had become united into a single continent. The collision zone was marked by a high mountain belt, whose remains can be seen today in Norway, the Scottish Highlands, and the northern Appalachians. Their origin is attested by deformation, metamorphism, and intrusion of granites produced by melting due to crustal thickening (see Chapter 4). This mountain building event or **orogeny**, is widely-known as the Caledonian Orogeny, though in the Appalachians it is usually called the Acadian Orogeny. Erosion in the Caledonian mountains shed sediment into rivers that deposited large amounts of sand in fresh-water basins in parts of the British Isles and eastern North America. These Devonian-age sandstones have long been known as the Old Red Sandstone, and the united Laurentia–Baltica continent is sometimes referred to as the Old Red Sandstone continent.

Battered though the Old Red Sandstone continent was along the multiple collision zone, its troubles were not yet over. In Carboniferous times the enormous continent of Gondwana hit it from the south, throwing up the fold mountain belt of the Ouachitas, reactivating and enhancing many of the Appalachian structures, and deforming much of western Europe and the north-west fringe of Africa. This orogeny is referred to as the Alleghenian Orogeny in America, and as the Hercynian or Variscan Orogeny in Europe and Africa. North of the new mountain belt, the deltas of rivers from the, by then, partly denuded Caledonian mountains provided the setting for widespread coal formation. To the south, Africa, South America, India, Australia and Antartic share glacial deposits of late Carboniferous and Permian age that attest both to their high latitude and the fact that these components of Gondwana were still united at that time.

In the Permian, Siberia collided with Baltica along the line of the present-day Ural mountains, which are a result of the associated mountain-building event. With the arrival of the rest of non-Gondwanan Asia shortly afterwards, virtually all the world's continental crust was united into a single 'supercontinent' known as Pangea (Figure 13.5). Europe (including the British Isles) and eastern North America now found themselves in low northerly latitudes. On land, desert conditions prevailed (as they do today at similar latitudes), and where the continental crust was flooded by shallow seas these were warm and poorly connected to the world ocean, so much of the deposition was in

Figure 13.5 The world in Triassic times (about 220 million years ago), when all the continents were united into the single supercontinent of Pangea. Tethys is the name given to the wedge-shaped ocean between the northern and southern parts of eastern Pangea. (Modified from various sources)

the form of evaporite deposits. Meanwhile volcanic activity above sub-duction zones and collision with island arcs continued to extend the continent in the present-day south-west of North America.

This was probably not the only time that all the continental crust was gathered together. It has been suggested that the situation illustrated in Figure 13.3 is a result of the breakup of a previous supercontinent that had been centred on Laurentia.

The break-up of the supercontinent of Pangea by continental rifting was as long and complicated as its construction had been. Much of the ocean floor that was created by these events still remains, especially in the Atlantic, Indian, and Antarctic oceans, so we can use the pat-tern of sea floor magnetic stripes (Chapter 4) to track the relative movements of the continents. Break up was initiated late in the Triassic, when rifting began between North America and Africa, but this did not progress far enough to generate oceanic crust until the Jurassic. Shortly afterwards, India broke free from between Africa and Australia and began to be carried northwards because of sea-floor spreading to its south and subduction beneath central Asia. Next, South America split from Africa, giving birth to the south Atlantic (Figure 13.6), but the north Atlantic between Greenland and Britain

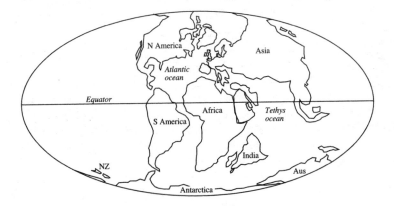

Figure 13.6 The world in late Cretaceous times (about 80 million years ago). The Atlantic had been opening between North America and Africa for about 100 mil-lion years, and between South America and Africa for about 30 million years, but had not yet begun to open between North America and Europe. India was begin-ning its rapid movement away from Africa and Antarctica, that would culminate in its collision with Asia about 40 million years later. (Modified from various sources)

did not start to open until about 65 million years ago, after a false start that began to split Greenland away from North America. Australia broke away from Antarctica at about the same time.

The lines along which Pangea broke apart appear to have been controlled at least partly by underlying hot-spots, where plumes rising from deep in the mantle caused heating, up-doming, and eventually rifting of the crust. Prior to rifting, extensive fields of basaltic lava flows were erupted, in the form of continental flood basalts (Chapter 5).

The westward movement of the Americas as Pangea broke up caused the oceanic plate of the Pacific to be subducted beneath them. This is the origin of the present volcanic belt from the Andes to Alaska, and helped to build the Central America land-bridge that currently links the two continents. As already noted, the history of western North America in particular is especially complicated because of the large number of exotic terranes that were accreted onto it.

Climate change

We have come across several examples of climate change in this book. When trying to use geology to interpret the past, it is important to realise that some variations in climate are local, related perhaps to the changing latitude of a drifting continent, or proximity to an ocean or sea, whereas others affect the whole globe.

Here we will just summarise the main global climate changes that can be deduced from the geological record. This is essentially a history of alternation between Ice Ages and warm periods. The globe appears to have been particularly cold during the late Proterozoic, with glacial features found even at low latitudes during most of the 250 million years leading up to the Cambrian. The global climate then warmed up for most of the Cambrian and Ordovician, followed by a 20-million-year Ice Age in the late Ordovician and early Silurian when glacial deposits are found in northern Africa (which was then at the south pole) and adjacent parts of Gondwana.

Warm conditions then returned for nearly a 100 million years until an 80-million-year Ice Age lasting from the early Carboniferous to the late Permian. Alternate advances and retreats of the ice and the associated falls and rises in sea level are probably responsible for the cyclic nature of sedimentation associated with the formation of coal deposits in late Carboniferous times, though undoubtedly subsidence of deltas and switching of delta channels played a role too (Chapter 9).

The world was mostly warm again from the late Permian until about the Eocene, when the climate began to cool. Antarctica was ice-covered for most of this time, but ice sheets did not spread from Greenland across North America and Europe until about two million years ago. As noted in Chapter 8, from the geological perspective there is no reason to suspect other than that our present mild climate represents merely a brief interglacial interlude between successive advances of the ice.

Concerns have recently been expressed that human activities may be forcing the climate to change at faster rates than most geological processes. For example, release of carbon dioxide by burning of fossil fuels is probably responsible for a slight, but perceptible, global warming since about 1950. This is because of the **greenhouse effect** in which carbon dioxide reduces the Earth's ability to radiate heat to space.

Another environmental condition that has fluctuated over geological time is sea level. Naturally, this has tended to be low at times when large amounts of water were held on land in the form of major ice sheets, and the highest when there have been no major continental ice sheets. However, an even more important factor seems to be the vigour of sea-floor spreading. When sea-floor spreading is particularly active, the total volume of constructive plate boundaries in the form of mid-ocean ridges is abnormally great. This displaces seawater unusually far over the edges of the continents, giving rise to atypically extensive shallow seas.

Sea level appears to have been as much as 600m higher than at present during the Ordovician, when sea-floor spreading following the break-up of the late Proterozoic supercontinent reached a crescendo. It had fallen to approximately its present level by the Triassic, when Pangea was in existence, but rose to a peak during the Cretaceous in response to the many young, hot, mid-ocean ridges initiated by Pangea's break-up.

Another factor that may affect the climate on a global scale is the distribution of the continents. For example in the Triassic (Figure 13.5) there was a single continental mass extending from pole to pole, which would have prevented the establishment of globe-encircling ocean currents. This is very different to the Devonian (Figure 13.4) when there was probably a globe-encircling ocean current running just south of the Equator, and to today, when a circumpolar current is able to circle Antarctica between 50 and 60°S. Another characteristic of today's climate is the monsoon, which brings much-needed seasonal rain to India and east Africa. According to the computer-assisted climate models, without the barrier to atmospheric circulation posed by the 5 km high plateau of Tibet, which is a result of the India–Asia collision, there would be no monsoon. Therefore, if there were no Tibet, the climate of the Indian ocean region, if not the globe, would be radically different.

The future history of the Earth

We can extrapolate the major elements of present-day plate motions with reasonable confidence for the next 50 or 100 million years. It seems likely that the Atlantic will continue to widen as sea-floor spreading on the Mid-Atlantic Ridge carries the Americas away from Europe and Africa. Eventually, a major subduction zone may form in the Atlantic, but we cannot predict on which side this will happen. Much of California seems set to continue its north-westwards displacement by sideways slip along the San Andreas fault system, and is due to pass British Columbia in about 50 million years time.

The collision of India with Asia is probably a spent force, and erosion will wear down the Himalayas. However they will be surpassed in grandeur by mountains thrown up by the forthcoming collision between Australia and south-east Asia. We are witnessing its first stages in the form of thrust movements in Papua New Guinea, and about 30 million years from now the whole of the Indonesia region is likely to be compressed into a continental collision zone. At the same time, slow northward convergence of Africa towards Europe is likely to unite the two continents, replacing the Mediterranean sea by a mountain belt higher and longer than the Alps.

Sea-floor spreading in the Red Sea will widen the gap between Africa and Arabia, and may propagate along a fault-line that runs along the present Jordan valley. Similarly, crustal stretching across the African Rift Valley may develop into sea-floor spreading, in which case a strip from Somalia to northern Mozambique will become separated by a new ocean from the main part of Africa.

Bounded on east, north, and west by subduction zones, the Pacific ocean will continue to be ringed by Andes-type volcanic belts and island arcs. If the pace of sea-floor spreading is unable to keep pace with the rates of subduction at the destructive plate boundaries, the Pacific ocean must contract. In this case, North America will eventually swing into mainland Asia, sweeping Japan to destruction in the process, and South America will collide with New Zealand and Australia.

If this happens, then, with the possible exception of Antarctica, the world in maybe 200 million years will once again bear a single supercontinent, though with its components differently arranged than in Pangea. Before long, this too will break apart, and the cycle will begin again.

Looking even further ahead, the Earth's radioactive heat engine is running down so slowly that there is no reason to expect that plate tectonics will cease to occur. The lithosphere may become a bit thicker, and sea-floor spreading a bit slower, but the large scale processes seem likely to continue much as they do today until (according to well established astronomical models) the Sun swells up and engulfs the Earth some 5000 million years into the future.

14

GEOLOGY ON OTHER PLANETS

One of the joys of studying geology is that when you have begun to understand the Earth you can use the same skills and knowledge to appreciate what has shaped the surfaces of other planets. On the other hand, interpreting pictures of other planets with the eyes of a geologist often gives new perspectives on the geology of the Earth.

We must exclude the giant planets Jupiter, Saturn, Uranus and Neptune, because any solid interior is hidden by a great depth of gaseous material. However, the 'terrestrial planets' Mercury, Venus and Mars are rocky bodies with (we presume) iron-rich cores that are directly comparable with the Earth. So is the Moon, which is a terrestrial planet too so far as geologists are concerned, despite the fact that it orbits the Earth.

There are a surprising number of other bodies in our solar system whose size and composition enable them to be studied geologically. These include the larger asteroids, Io (a rocky satellite of Jupiter with active volcanoes), and the nineteen largest icy bodies in the outer solar system. The latter have an icy mantle overlying a rocky core, but the behaviour of this ice mimics in all important respects the behaviour of the silicate material that constitutes the Earth's outer layers. Most of these icy bodies are satellites of the giant planets, but the planet Pluto and its satellite Charon belong to the same class.

In this chapter we will review the geology of solar system bodies in turn, though there is not space to do any of them justice. However, first it is worth saying something about impact cratering. This is the clearest example of a process whose effects are studied more easily on other bodies, and whose study has led to better insights into the Earth itself.

Impact craters

Any solid surface in the solar system that survives erosion, burial (by sediments, lava or volcanic ash), or deformation by tectonic processes for a few hundred million years becomes pockmarked by circular craters. The Moon is the most obvious example, its largest craters being visible even with binoculars. Counting the density of craters on planetary surfaces is the best way we have of estimating surface ages.

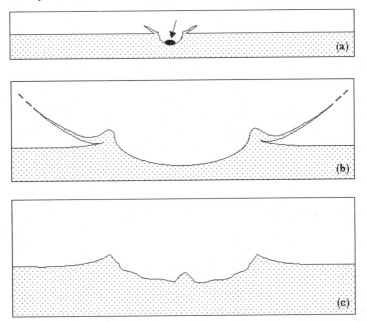

Figure 14.1 Cross-sections to show how an impact crater is formed. (a) A high-velocity projectile strikes the surface. A shock wave begins to excavate a crater. It also destroys most of the projectile. (b) Excavation by the shock wave proceeds radially out from the point of impact. As drawn at this stage the crater has reached virtually its final diameter and the new ejecta is barely able to flop over the rim. (c) A central peak has risen by rebound, and the inner walls of the crater have slumped, giving the crater its final form. Where there is no atmosphere to shield the surface, impact cratering occurs at all scales from micrometres to 1000 km. A single central peak, as shown here, occurs only across a limited size-range (15-140 km on the Moon, 7-70 km on the Earth). It takes only about 100 seconds to excavate a 100 km diameter crater.

Comparative studies of craters on different bodies and laboratory simulations of crater formation have given us a very clear idea of how craters form. Figure 14.1 shows the accepted model. A projectile (such as a fragment of an asteroid) hits the surface at about 10 to 40 km per second. At such a high velocity, crater formation is driven by shock waves generated at the point of impact. There is no need for the projectile to hit the surface at right angles for the crater to be circular. The shock waves excavate a hole and fling out a 'curtain' of fragmented ejecta that is dispersed over the surrounding terrain. The crater ends up about thirty times the diameter of the projectile. Impact craters have a distinctive and virtually unmistakable morphology, which includes a central peak whose presence, size and form is consistent with the crater diameter, a floor that is lower that the surrounding terrain, a circular outline, and a surrounding blanket of ejecta.

Studies of impact craters on the Earth confirm the high pressures associated with the shock of impact. The first such crater to be thoroughly investigated was the Barringer crater, otherwise known as Meteor Crater, in Arizona. This is a small crater (barely a kilometre across) that was formed about 20 to 30 thousand years ago, but it has a characteristic ejecta blanket around it, and drill holes show that the rock below the crater floor is severely shocked.

Tremendously high impact pressures are also demonstrated by the 15-million-year-old 24-km Ries crater near Nördlingen in Bavaria, Germany. This is strewn with diamonds thought to have been produced by shock compression of atmospheric carbon dioxide. Unfortunately for the good people of Nördlingen, these diamonds are microscopic in size and of no conceivable use. However, as we saw in Chapter 9, meteorite impacts have sometimes provided genuine bounty from the skies as exemplified by the Sudbury nickel ore body in Ontario.

The dangers posed by impacts are most clearly demonstrated by the 'smoking gun' of the Chicxulub crater in Mexico whose formation led to the end-Cretaceous mass extinction (Chapter 12). Even an impact as small as the one that produced the Barringer crater would destroy any nearby city. The best estimate is that something of this sort can be expected, on average, once in about 50 thousand years. The ongoing risks, and lack of warning, of such impacts were dramatically demonstrated during the writing of this chapter when a 300-metre-wide asteroid (1996 JA1) was discovered on a Tuesday and passed the Earth at a distance of only 450 thousand km on the following Sunday.

That is just a bit greater than the Earth–Moon distance, and is a 'near miss' by solar system standards.

The Moon

The Moon is effectively the smallest of the terrestrial planets. Its diameter is 25%, its mass 1.2%, and its density 60% that of the Earth. It is too small to retain an atmosphere. Because it is so close, the Moon has been explored more fully than any body other than the Earth. A calibrated cratering timescale has been established from the relationship between radiometric ages of samples of lunar material and the local density of impact craters. By extrapolating this, we can make a reasonable estimate of the absolute ages of cratered regions on other bodies in the inner solar system.

The Moon appears to have cooled down a lot quicker than the Earth. This is to be expected, because a planetary body loses heat in proportion to its surface area, but generates heat (by radioactive decay) in proportion to its volume. The smaller the body, the greater the amount of surface area relative to its volume, so the faster it will cool. In the case of the Moon, this cooling has produced a rigid lithosphere about a thousand km thick, which explains why the lunar surface shows no signs of current or recent geological activity.

The Moon's geological processes waned so quickly that, as noted in the previous chapter, large regions of primary crust have survived, in the form of the pale, anorthosite, highland terrain, dating back about 4400 million years. The dark areas on the Moon are covered by basaltic lavas that flooded most of the largest impact basins between about 3900 and 3100 million years ago. These dark areas were once (incorrectly) thought to be lunar seas, and they are still known by the Latin word for sea 'mare' (pronounced mah-ray; the plural form is 'maria'). A rapid decline in the rate of meteorite bombardment in the early solar system is demonstrated by the fact that the lunar highlands typically have at least ten times as many craters per unit area than the maria, despite being only about 25% older.

The craters in the lunar highlands show virtually no signs of deformation by tectonic processes, so the Moon's lithosphere must have become thicker (or more rigid) than the Earth's at an early stage. After the most recent episode of widespread mare lava eruption, it evidently became too thick for more magma to escape from below and so far as we can tell the Moon has been geologically dead (except for impact cratering) ever since.

Mercury

Mercury is the closest planet to the Sun. Its diameter is 38%, its mass 5.25%, and its density 98% that of the Earth, and it lacks an appreciable atmosphere. It probably has a much larger core, relative to its total size, than the Earth. The spaceprobe Mariner-10 sent back images in 1974-75 showing that Mercury is a heavily-cratered world, in many ways reminiscent of the lunar highlands. It has clearly been geologically dead for most of the age of the solar system, but there are enough hints of volcanic processes and tectonic features to make it worth going back for more detailed look.

Venus

In terms of basic properties, Venus is our twin planet. Its diameter is 95%, its mass 82%, and its density 95% of the Earth's. It is shrouded beneath an opaque atmosphere, but orbiting spaceprobes have been able to image the surface using radar. Venus's atmosphere is denser than ours, with a surface pressure over 90 times greater, and is 96% carbon dioxide. Most of the Earth's carbon dioxide is held in the form of limestone, but this can only form in water and Venus is too hot for liquid water to exist. The large amount of carbon dioxide in Venus's atmosphere has caused a greenhouse effect far more severe than that proposed for the Earth in the previous chapter, to the extent that Venus's surface temperature (about 450°C) has become much hotter than it would otherwise have been.

With virtually the whole globe revealed by radar images in sufficient detail to show features as small as about a hundred metres across, the surface of Venus is better documented than the most of Earth's ocean floor. However, our understanding of Venus does not compare with that of the Earth, because we lack seismic data and other information that can only be gathered actually on the ground.

Perhaps the best way to sum up Venus is to remark that, contrary to expectations, it does *not* appear to be experiencing Earth-style plate tectonics. There are widespread volcanoes and lava flows that appear to be basalt, and there are some highly deformed regions (Figure

14.2), but nobody has come up with a convincing model that is capable of linking the various elements together on a global scale in the way that plate tectonics does for the Earth.

Figure 14.2 A radar image showing a 300 km wide area on Venus. A belt of fold mountains, similar to the Appalachians, runs from east to west across the picture.

A particularly curious aspect of Venus is the age of its surface. Crater counts show no significant variation from region to region, and the most straightforward explanation of this is that the whole globe was resurfaced within the space of a few tens of millions of years approximately 500–800 million years ago. Some of the volcanoes may have been active considerably more recently, but only on a small and localised scale.

The apparent lack of on-going plate recycling and recent volcanism on Venus indicates that the current rate of heat transport from below its lithosphere is very low. This is surprising, because radioactive decay in Venus's interior ought to be generating heat at a rate not much less than within the Earth. If we accept this, then we are drawn to the conclusion that most of this heat is being trapped *below* the lithosphere. The asthenosphere must therefore be warming up, and (because of thermal expansion) becoming more buoyant, so the dense lithosphere will eventually founder, and magma from the asthenosphere will spill out over the surface. This will be another global resurfacing event, similar to the one inferred 500–800 million years ago.

If this model for Venus is correct, then its behaviour is spasmodic. Most of the heat-loss occurs in relatively short episodes of intense activity, separated by long periods of quiescence. This contrasts with the Earth, where heat is lost steadily by the creation of new hot lithosphere at constructive plate boundaries and the subduction of old cold lithosphere at destructive plate boundaries. Why the two planets should behave so differently is one of the major mysteries of planetary science, and we are left wondering if the young Earth, say during the Hadean or Archean, behaved as Venus does today.

Mars

Mars is smaller than Venus, but larger than Mercury. Its diameter is 53%, its mass 11%, and its density 72% of the Earth's. Like Venus, its atmosphere is mostly carbon dioxide, but the surface pressure is less than a hundredth of the Earth's.

Mars appears to have the cool, thick lithosphere that we would expect in view of its small size. It has several impressive volcanoes, including the largest in the solar system. This is Olympus Mons whose summit is some 24 km above its base, and it would dwarf the Big Island of Hawaii, which is the Earth's largest volcanic edifice (Figure 14.3). The vast bulk of Olympus Mons could be supported only by a thick, strong lithosphere. Moreover, the concentration of volcanic activity at this point over many hundreds of millions of years indicates that Mars's lithosphere is stationary relative to the hot plume within the mantle that presumably feeds the volcano.

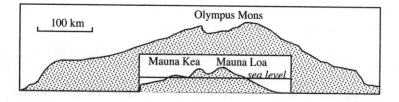

Figure 14.3 Cross-sections to show comparative sizes of Olympus Mons on Mars and the Big Island of Hawaii (its two largest volcanoes, Mauna Loa and Mauna Kea are named). The horizontal scale is as shown, but the vertical scale is exaggerated fivefold.

Large though it is, Olympus Mons covers too small an area for impact crater statistics to give a reliable estimate of its age. The best guess is that it last erupted about 30 million years ago. There are several smaller (but still very large) volcanoes nearby and others elsewhere on the globe.

For a smallish planet, Mars has a surprising number of exceptional features in addition to its volcanoes. The most fundamental is a two-fold division into an ancient highly-cratered highland terrain occupying most of the southern hemisphere and a younger low-lying terrain in the north. The highland terrain has many channels upon it, some of them in dendritic networks (e.g. Figure 8.4a) which can hardly have been cut by anything other than flowing water. The younger surface age of the northern lowlands is probably because of deposition of sediment transported from the highlands. It has even been suggested that part of the northern hemisphere was once occupied by a shallow sea. That being so, it would be rather surprising if primitive life had not developed, as indeed is indicated by traces found in rare meteorites believed to have come from Mars.

Liquid water could not exist under the low atmospheric pressure that prevails on Mars now, so clearly its climate must formerly have been very different. Mars has lost much of its atmosphere and cooled down over time, in contrast to Venus which has warmed up and developed a dense atmosphere.

Based on ages inferred from crater counting, liquid water seems to have disappeared from Mars about 2000 million years ago. Water still occurs in the form of ice in the northern and southern polar caps. In winter these are enlarged by a seasonal covering of frozen carbon dioxide. In the absence of water, sediment transport and erosion of surface features are today restricted to wind effects. Extensive sand dune fields occur, and dust storms sometimes obscure the surface for weeks on end.

Mars is also notable for Valles Marineris, a 4000-km-long canyon system in the southern highlands that reaches 7 km in depth and over 600 km in width. This would dwarf the Grand Canyon even more than Olympus Mons would overshadow Hawaii. It probably owes its origin to tectonic fracturing, but has been extensively modified by flowing water and landslides.

The asteroids

As noted in Chapter 2, the asteroids, mostly orbiting the Sun between Mars and Jupiter, are probably the remnants of planetesimals that would have grown into a planet but for the gravitational perturbations caused by Jupiter. Most are rocky bodies, though some appear to be metallic (like the Earth's core). The largest of them, Ceres, is about 900 km across, which is far too small for radioactive heat production today to keep its interior warm enough to be mobile. All the asteroids are thus lithospheric throughout, and are not expected to display evidence of recent internally-driven geological processes.

However, images of Vesta (the third largest asteroid, 500 km in diameter) obtained by the Hubble Space Telescope are clear enough to reveal large dark patches on its surface. These could be Vesta's dark interior, exposed at the surface as a result of collisions, in which case Vesta would seem to have a compositionally differentiated structure, analogous to the crust-mantle distinction in the Earth. Alternatively, the dark patches could be lava flows. Either interpretation calls for Vesta's interior to have been hot at one stage. If this is so, then the explanation probably lies with short-lived radioactive isotopes, such as aluminium-26 (^{26}Al), which we have good reason to believe could have been very significant heat sources during the solar system's first hundred million years or so.

One of the few asteroids to have been visited is 55-km-long Ida, which the spaceprobe Galileo flew past in 1993 on its way to Jupiter. Ida is irregular in shape and covered in impact craters. This was to be expected, but most scientists were surprised to discover a tiny asteroid less than 2 km across in orbit around it, which was subsequently named Dactyl.

Io

Io is the innermost of Jupiter's four large satellites. By analogy with the Moon, which it closely resembles in size, density and inferred composition, one would expect Io to be a densely cratered body with a few signs of geological activity in the past 3000 million years.

The reality is very different, because Io is the most volcanically active body in the solar system. Up to about ten volcanoes are erupting at any one time, and the surface is being covered by volcanic material at a rate more than adequate to have buried the old impact craters.

There are both explosive eruptions, in the form of 300-km-high eruption plumes, and lava flows. When first seen, on images from the Voyager spaceprobes in 1979, the lava flows were widely interpreted as sulfur because of the red or yellow colour of Io's surface. Since then, thermal infra-red studies using telescopes have demonstrated that the temperatures in the volcanic vents are too high for molten sulfur, and it seems that we are dealing with silicate lava (probably basalt) of the kind familiar on the Earth. Io's colour is almost certainly caused by sulfur, but it is merely a surface dusting, as is sometimes seen on lava flows on the Earth.

Radioactive heating cannot account for Io's prodigious volcanism. The explanation appears to be tidal heating of Io's interior, caused by tidal deformation of Io's shape as it orbits Jupiter.

Large icy bodies

Io is the outermost of the solar system's major rocky bodies. However, Jupiter has three other large satellites. The largest of these, Ganymede, is bigger than the planet Mercury. However its density is much less because it consists of about three parts ice and one part rock (Figure 14.4). Saturn has one large icy satellite, Titan, that rivals Ganymede in size, and six others more than 400 km across, Uranus has five satellites larger than this, Neptune has two, of which Triton (2700 km across) is the largest and most interesting. The solar system's inventory of the large icy bodies is completed by the planet Pluto (2280 km diameter) and its satellite Charon (1200 km diameter).

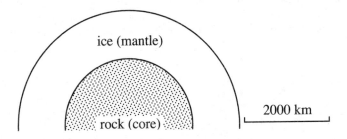

Figure 14.4 Cross-section through an icy body, showing differentiation into a dense rocky core surrounded by less dense ice. The scale bar is relevant to Ganymede (which is now believed to have an inner core of iron), but many smaller icy satellites, and also the planet Pluto, probably have similar structure.

We can ignore the forty or so other known satellites of the outer planets, and various icy objects further from the Sun than Pluto, that are less than about 400 km across, because below this size the body's own gravity is too weak to pull it into a spherical shape. Unlike some of the asteroids, we have no reason to suspect internal differentiation or other geological processes in such irregular bodies.

It may seem strange to deal with icy bodies in a book about geology, but their relevance becomes clear if we review the properties of ice. At the low surface temperatures prevailing in the outer solar system (-150°C at Jupiter and -240°C at Neptune) ice is extremely strong and rigid. However, in the interior of a largish body with a supply of heat, ice is capable of flowing by convection without melting. This distinction between rigid and convecting behaviour is exactly the distinction described in Chapter 2 between the lithosphere and asthenosphere of the Earth (or, indeed, any other terrestrial planet), except that we are dealing with ice rather than silicates.

A further parallel with silicates arises when we consider how ice melts. Unless it is pure, ice does not melt at a single temperature. Salty ice, such as we might expect in Jupiter's icy satellites because of reactions between water and rock, melts at about 4° below 0°C, yielding a brine instead of pure water. Ice that consists of a mixture of frozen water and frozen ammonia, as is likely at Uranus and beyond, melts at about 100° below 0°C, producing a liquid that is about two parts water and one part ammonia. This is partial melting, just like that occurring in the mixtures of silicate minerals that make up rock and described in Chapter 4.

Thus with a lithosphere/asthenosphere distinction and the potential to produce compositionally distinct 'magma' by partial melting, icy bodies have the essential attributes of a terrestrial planet. However, even the large ones would be too cold inside for this to matter, unless tidal heating kept their interiors warm. Fortunately for the geologist, most of them show plenty of evidence for current or previous episodes of tidal heating. For example parts of Ganymede's surface are fractured on a very fine scale (Figure 14.5), but the large number of impact craters superimposed on this pattern shows that it is preserved from the distant past. On the other hand another of Jupiter's satellites, Europa, shows many signs of much more recent activity including both fracturing and eruption of watery 'lava' flows. Europa's icy mantle is only about 100 km thick, and its lower part is probably

Figure 14.5 A highly detailed image of a 35 km wide area of the icy surface of Ganymede (a satellite of Jupiter 5260 km in diameter). The cross-cutting pattern of ridges and grooves reveals that Ganymede's surface has experienced a complex history of fault movements and resurfacing by icy volcanism. The small circular features are impact craters.

liquid water overlying tidally heated rock where there could be chemically rich hot vents like those where life may have started on Earth.

Moving out to the satellites of Uranus, we find features such as 'lava' flows sketched in Figure 14.6, which were presumably composed of an ammonia-water mixture produced by partial melting. It is known that such a mixture is vastly more viscous than pure water, and would behave very much like a silicate lava flow on Earth.

Figure 14.6 A sketch made from an oblique view of part of Miranda (a 470 km diameter satellite of Uranus) with a particularly complex geological history. Several impact craters are visible (roughly circular features). One 10 km crater in the lower left has been partly obliterated by a resurfacing event, whereas a similar crater nearby evidently post-dates this as it is superimposed across the boundary. Two icy 'lava' flows are shown in bold outline.

Geology among the stars

Planets have now been discovered around other stars. So far the discoveries have been roughly Jupiter-size or bigger, because these are the easiest to detect, but Earth-like planets must exist around such stars too. In our lifetimes we should get spectroscopic data on their atmospheric compositions, but we cannot expect to see their surfaces in sufficient detail for any kind of geological study. However, with many unsolved questions remaining about the Earth, and the geological exploration of our solar system barely begun, there is plenty to keep the geologist busy, as well as fascinated, nearer to home.

15

GOING INTO
THE FIELD

Using spacecraft to explore the surfaces of other planetary bodies is both fun and informative, but there can be no denying that for most geologists, whether professionals or hobbyists, the main attraction of geology is doing it in the field. Fieldwork can be a great source of pleasure, but it is also the best way to appreciate the sort of evidence from which most of our understanding of the Earth's history is derived.

A government administrator once asked me 'Where exactly are these fields that you work in?' Perhaps the term 'fieldwork' is another of those pieces of geological jargon that seems designed to confuse the layperson. However in this case at least it's a term that we have in common with botanists, zoologists, ecologists and even sociologists. When a geologist speaks of 'going into the field', he or she means spending time outdoors in a place where the rock sequence to be investigated can be seen. Thus, fieldwork can be conducted in places as diverse as a sandy beach, a roadside cutting, a remote stretch of moorland, or the flanks of an active volcano.

If you belong to a geology club, or are studying a course in geology, you have probably been on at least one field trip by now. However, if you have never been in the field, and have decided after getting to this stage in the book that you want to go deeper into geology, then I urge you to get out there and begin looking at some rocks. This chapter suggests some ways to begin practical geology in the field. It therefore describes things to look out for, discusses the equipment you will find useful, and suggests a code of practice to enable you to work safely and without annoying any landowners.

Things to do in the field

It is sensible to decide in advance what you want to achieve during a field trip. For example, are you just going out for a general look at the rocks, or do you want to search for the answer to a particular riddle? If you are going in a group organised by somebody else you can leave the decisions about where to go and what to look at to them, but to get the most out of a field trip it pays to approach each locality in a methodical fashion.

When you arrive at a rock exposure, perhaps a cliff face, do not rush straight up to it. I will have something to say about the safety of working under cliffs later, but it makes scientific sense to stand back for a while anyway.

You should spend time taking in the general features of your site. These are best seen from a distance. The sort of thing to look out for is whether or not there is any layering in the rock, which might take the form of bedding, layering in an igneous intrusion, metamorphic foliation, or a set of fractures (joints). You may not be able to decide which of these four you are seeing, but that is not the point at issue for now; you will be able to tell for sure when you get close up.

While you are still ten or more metres away, you should look for variations. For example, if you can see some layering try to decide whether these layers maintain a uniform orientation throughout the exposure, or whether they are folded.

You should also visually trace some layers sideways to see if they are continuous. If they are not, try to decide whether this is an inherent property of the layers (perhaps each one represents a deposit laid down in a channel that had a finite width, or maybe you are seeing large-scale cross-bedding of the sort that occurs in wind-blow sand dunes), or whether originally continuous layers have been truncated by an igneous intrusion or by a fault (which will almost certainly off-set the layers).

If the structure within an exposure is in any way complicated you can often begin to sort things out if you make a simple sketch. This should not be a piece of representational art, showing every crack, shadow and tonal variation. Rather, it should show *only* those features that are of geological significance. For example, bedding is significant and should be drawn, but if you try also to show all the irregularities in a rock face left by either natural weathering or quarrying then you will obscure the essential features. Always include a scale in a field sketch. As a rough guide, most adults are 1.6–1.8 metres tall, so if you can persuade somebody to stand against your exposure you can very easily mark a scale-bar on your sketch. A sketch is usually more useful than a photograph, because it shows only what you need and you can annotate it while you are in the field.

Before you finish your preliminary survey, pick a suitable place to begin your close-up examination. Avoid approaching anywhere you are likely to be hit by material falling from above, or where the ground beneath you could be unstable. Furthermore, do not head for somewhere that looks atypical. Instead, choose a spot that looks representative of the exposure in general. It is best to try and make sense of any strange bits *after* you have worked out the main story.

By the time you arrive at the rock face, your preliminary observations will have suggested a number of checks needed to test any hypotheses that have begun to form in you mind. If you had decided that this exposure was probably sedimentary rock, you would expect to be able to confirm it by close inspection, which should reveal the individual grains making up the sediment (unless these are very fine), small-scale bedforms, and (if you are lucky) trace fossils and/or body fossils. On the other hand, a coarse-grained igneous rock would manifest itself by clearly visible interlocking crystals, including minerals of the sort that do not survive transport in the sedimentary regime (Chapter 8). In order to identify the rock, you would need to recognise and identify its main minerals, using Appendices 1 and 2 or a more detailed field guide to rocks and minerals to help you.

You may see features close up that are worth sketching or photographing. Examples would be small-scale cross-bedding, fossils, or the shape of the contact at the edge of an igneous intrusion. Just as with any more distant sketch (or photograph), you should try always to include an object that will give a clue as to scale: a coin, a lens cap, or a colleague's hand are probably the most common scale objects. The latter has the advantage of not being liable to be accidentally left behind when you leave.

If you are not visiting a quarry or a cliff, you are likely to be dealing with comparatively small exposures, unless you are in an arid environment where there is little or no soil cover. However, the principles outlined above still apply. Stand back and assess the setting of your exposure, however small, before rushing in and pressing your nose against it.

Field equipment

It is a common misconception that geologists spend their time digging up rocks and fossils. It is rare for a geologist to go into the field armed with a spade. The trick is to go somewhere where the things you want to see are already at the surface. In many cases, the natural weathering of the rock picks out features more clearly than they can be seen on freshly exposed surfaces. Exposures of rock are most common in hilly or mountainous regions, especially along water courses, but they may be particularly good on the coast. Quite apart from the damage and disturbance that digging would cause, the labour involved in removing two or three feet of topsoil would rarely be repaid by the small patch of dirty and possibly frost-shattered bedrock that this would reveal.

The most important items of field equipment are a perceptive eye and an enquiring mind. There are no artificial aids for the latter, but the former can be augmented by carrying a hand lens. A typical hand lens is smaller than a Sherlock Holmes-style magnifying glass, but provides a magnification of about ten times. This can be extremely helpful for such things as determining the shapes (angular or rounded) and textures (frosted or glassy) of sedimentary grains, spotting cleavage traces in minerals, and making out fine details in fossils.

You should also carry a notebook, for recording your observations and making sketches. Pencil is best because it can be rubbed out to correct mistakes, and does not smudge as badly as ink when it rains. If you might want to go back on another day to check or extend your observations it is important to record you exact location. A grid reference on a map and/or a sketch map of your own should do the trick.

Sometimes you may want to collect a small sample. If you are just starting out as a geologist it is an especially good idea to build up your own collection of the main types of rocks and fossils, but see my later warning about selfish and indiscriminant collecting. If you do find something that you want to take away, then you need something to put it in, and some way of labelling it. One specimen per small plastic bag, each containing a sample number on a slip of paper is a good way to proceed, provided you remember to note the exact location of each find in your notebook.

If not with spades, then it is with hammers that geologists are usually associated. Unlike spades, hammers are a standard item in many geologists' field gear. In situations where the rock surface is too badly weathered or covered by lichen for it to be identifiable, a hammer can be invaluable for breaking a piece off to expose a fresh surface. However, natural freshly broken pieces can very often be collected near the foot of cliffs, in which case there is no need to hammer the rock face at all. The same goes for collecting fossils; you will usually find many more, and avoid damage, by searching through rock falls or spoil heaps. If you do use a hammer, then it must be a purpose-made geological hammer. The head of a hammer sold for knocking in nails will not be made of adequately tempered steel, and is liable to splinter if struck against a rock. A steel shard from a shattered hammer can cause severe lacerations, and may blind you if it hits you in the eye.

The other standard item of field equipment is a device called a compass clinometer, which is used to measure the orientation of bedding, foliation, fault-planes and so on. This is particularly important if you are being ambitious enough to draw your own geological map. It is valuable too when, for example, you are visiting several isolated exposures of a sedimentary rock and want to determine if the bedding in each one has the same orientation. If it does not, then the rock sequence has probably been folded.

A compass clinometer has a magnetic needle, functioning as in any ordinary magnetic compass, and a second, non-magnetic, needle that is pivoted from one end so as to hang vertically under its own weight. To appreciate how to use a compass clinometer, it is necessary to understand how the orientation of any plane surface can be specified in a unique and unambiguous manner. This can be done with just two measurements. The first is to imagine a horizontal plane intersecting the geological surface (such as a bedding plane) whose orientation we want to determine. The two planes must intersect at a horizontal line and the angle between this line and north is defined as the **strike** of our geological surface (see Figure 15.1). To measure the strike, the compass clinometer is held horizontally, with one edge pointing along the strike direction. The angle that this edge makes with the free-floating magnetic compass needle is the strike, and can be read off the dial. The only other measurement we need to know is how steeply the geological surface is dipping, relative to horizontal. This is called the **dip** of the surface. To determine this, the compass clinometer is held with one edge running directly down the steepest gradient that can be found on the geological surface. The dip can then be read using the free-hanging needle.

GEOLOGY

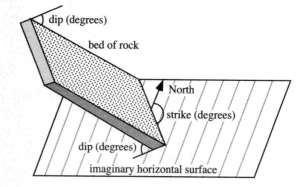

Figure 15.1 Definitions of strike and dip. A bed of rock is shown, but the definitions apply to any geological feature whose local orientation can be expressed as
a plane. Strike is the angle, relative to magnetic north, made by any horizontal
line on this geological surface. Dip is a measure of how steep the surface is, and
is defined as the angle of its slope measured downwards from horizontal.

In the example shown in Figure 15.1, the strike is 150° and the dip is
35°. This would be recorded in our field notebook as 150/35. This is
unambiguous if strike is always measured clockwise from north (as in
this example), but to play safe it is common to note down the direction
in which the surface is dipping. In this example it is dipping down
towards the north-east, so we could record 150/35NE. Figure 15.2
shows a compass clinometer in use.

If you want to record your visit photographically, it is best to have a
camera with adjustable focus, or even a choice of lenses. Some landscapes are best recorded using a wide angle lens, but detailed close-up
photographs require a macro lens.

Other items of field equipment depend on the nature of the rocks. If it
is in a poorly-cemented sandstone, you may find a trowel or similar
object handy for scraping away loose grains in order to reveal the
original cross-bedding. If it is a shale and you hope to find fossils, a
chisel comes in handy for splitting the rock along its bedding surfaces

Finally, if there is any likelihood that your field trip will involve
working close to steep or vertical faces more than about three metres
high then you should make sure that you wear a hard hat. These are
cheap and readily obtainable from builders' merchants and do-it-your-
self stores.

Figure 15.2 A compass clinometer being used to measure the strike of the edge of a pale igneous dyke, which has been intruded into a darker rock. The photograph was taken looking downwards onto a roughly horizontal surface. In this example the strike is 160° relative to magnetic north. The edge of the dyke dips at 80° towards the east (left in this view), so the strike and dip of the dyke margin would be recorded as 160/80E.

Field safety

In the field, you should take care to minimise the likelihood of injury to yourself and to others. As already noted, the most basic precaution is to assess the inherent risks before approaching a rock face. You should wear your hard hat whenever there is the slightest chance of anything gravel sized or above falling onto you. While it won't help you survive a major rock fall, a hard hat can save you from concussion by small lumps of debris.

If it is absolutely necessary to use a hammer, then you must make sure that nobody is within about three metres of you. Rocks such as flint and many igneous varieties are particularly hard and brittle and hammering can send sharp splinters flying. Most older geologists have met someone who has lost the sight of an eye in this way, and it is foolish to hammer without wearing goggles or safety spectacles.

Be adequately clad. Walking boots with ankle protection are the most sensible footwear if you will be working in or crossing rough terrain, and you should carry protection against sun, rain and cold. When visiting a coastal exposure, be aware of the state of the tide and plan ahead to avoid being cut off. When visiting a remote area, general back-country rules apply: be wary of dangers such as being caught in a gully or canyon in the event of a flash flood, make sure you have enough food and drink to survive if your return is delayed for some reason, and leave a copy of your intended route with someone in case you have to be rescued.

Fieldwork code of practice

The rules governing geological collecting and other activities vary from place to place. However, there are various universally applicable guidelines.

- If you are entering a quarry or other private land, seek permission. Keep to paths, close gates behind you, and do not climb over walls or fences. Quarry owners will usually insist that you wear a hard hat at all times, and that you stay well clear of machinery.

- If you are working in a national or state park, nature reserve, or other conservation area, then appraise yourself of the rules about collecting. Some places will allow you to collect samples only from loose debris, others ban collecting entirely.

- Even where there are no local rules, be aware of the damage you might cause. Hammer, if at all, with discretion. Do not damage rock faces, and do not try to extract fossils or mineral specimens that are well embedded. You will usually end up wrecking them, so it is much more sensible to take a photograph, and leave the fossil or mineral for others to admire.

Geological maps and guidebooks

If you want to know about the geology of a particular area, then an obvious first step is to buy a geological map. A good bookshop will be able to order these for you. A geological map usually shows the distribution of rocks that would be seen at the surface if the superficial deposits (such as soil) were removed. Sedimentary rocks are usually colour-coded by age, whereas igneous and metamorphic rocks are distinguished by type. Major faults are usually shown by bold lines.

Most geological maps are published by geological survey organisations. The geology of the United Kingdom is conveniently covered in two sheets at a scale of ten miles to the inch published by the British Geological Survey. Most of the country is available on larger scale maps, 1:50 000 being the commonest, which include additional information such as strike and dip, mineral veins, and so on. The equivalent organisation in the USA is the United States Geological Survey, but many state geological surveys publish maps too.

A map will show you the distribution of rock types, but it will not usually tell you where these are well exposed or conveniently (and legally) accessible. This is where geological guidebooks come in handy. They usually contain sketch geological maps too, so for beginners, at least, there is often no need to invest in a large scale geological map. Many guidebooks are written by enthusiastic geologists who know and love the area, others are more in the nature of official guides round geological trails in national parks.

Suggestions for field projects

Here are a few suggestions for geology fieldwork projects, that you could tackle after reading this book. There will probably be one or two that can conveniently be attempted in your home area or on your next holiday.

- If there are quarries in your area, find out whether these occur in every local rock type. If they do not, then try to work out why some rock types are (or were) quarried and others not.

- Visit a river or stream, and decide whether it is mainly eroding or depositing sediment. Does it do the same thing all year round? Could it be described as a braided or a meandering channel? Can you find any abandoned meanders (like the ox-bow lake in Figure 9.7)?

- Take your hand lens to the seaside. If it's a sandy beach, you should be able to identify quartz grains (clear, glassy) and shell fragments (milky, opaque) among the sand particles. What else can you see? Where so you think it has come from? If it's a pebbly beach, how many different rock types are represented among the pebbles? Are they all local? Do they match the rock types exposed in nearby cliffs?

• Find an exposure with fossils in it, and try to identify them and work out how they lived (swimming, fastened to the sea floor, in burrows, etc.). Are they all the same kind? Are most of the fossils complete or fragmentary? What can you deduce about the environment in which the fossils (i) lived, (ii) died? Are (i) and (ii) the same?

• Is there any history of mining in your area? What is or was extracted, and how was it formed? Can you find any signs of subsidence?

• What are the sources of the stone used (i) to build houses, (ii) to build churches, colleges, and municipal buildings, (iii) to decorate the facades of banks, department stores and so on, (iv) for gravestones? Is it all local, or has some been brought in from tens or hundreds of miles away? Gravestones usually bear dates; do some varieties of stone weather differently from others, and if so why?

• Visit some really big hills or mountains. Why are they there? How many rock types are present? Can you find any evidence of deformation (such as folding or cleavage)? Has the landscape been eroded into its present form mainly by the action of flowing water or by ice?

The possibilities for field projects are endless, and you can probably think of ideas that appeal to you more than those I have suggested. Best of all, join a local geological society, which will probably offer a programme of talks and field trips.

APPENDIX 1: MINERALS

There are many thousands of minerals known. Guides to the identification of even the few that are relatively common take up many pages of colour photographs and lists of properties. This appendix is not intended as a substitute for such a guide; rather it is a brief introduction that indicates some of the ways in which the very commonest minerals may be identified.

It is not straightforward to define what is meant by the term 'mineral'. A simple statement that 'a mineral is a naturally occurring crystalline substance with a well defined chemical composition' is a good starting point. However, this is too restrictive to be an adequate definition, because the exact chemical composition of a mineral may vary, although the arrangement of atoms within its crystalline structure is fixed. On the other hand, some minerals that are regarded as distinct from each other share the same atomic arrangement, and are distinguished by differences in their chemical composition (which may radically affect their colour or other properties). Some minerals have *identical* chemical composition but the atoms are arranged differently giving the crystals totally different properties (these are described as 'polymorphs'). A few substances of non-crystalline form, notably amorphous (or 'cryptocrystalline') silica such as chalcedony, flint and opal, are usually regarded as minerals too.

A thorough analysis of a mineral would require chemical analysis, study of its optical properties using a microscope, and X-ray analysis of its atomic structure. However, a common mineral occurring as a crystal bigger than a few millimetres long often shows enough clues for the moderately experienced geologist to make a fairly confident identification. There are several things to look out for. These include: lustre (does it look metallic or glassy?), colour (but impurities can make this misleading), cleavage (how many sets of cleavage planes, if any, and at what angle do they intersect?), general shape (related to cleavage in some, but not all, minerals), hardness (how easily can it be scratched?).

The hardness of minerals is judged on Mohs' scale of hardness, which runs from 1 (softest) to 10 (hardest). A harder mineral is able to scratch a softer one. Common minerals are used to define each point on the scale as show in the table below:

1	Talc	can be scratched by fingernail
2	Gypsum	can be scratched by fingernail
3	Calcite	can be scratched by copper coin
4	Fluorite	can be scratched by steel
5	Apatite	can be scratched by steel
6	Feldspar	scratches glass with difficulty
7	Quartz	scratches glass with ease
8	Topaz	scratches glass with ease
9	Corundum	scratches glass with ease
10	Diamond	scratches glass with ease

Table A Mohs' scale of hardness.

The most common rock-forming minerals, notable ore minerals, and various precious and semiprecious stones are listed in the tables below:

Name	Hardness	Distinguishing properties
Amphibole (e.g. hornblende) $(Na,K)Ca_2(Mg,Fe,Al)_5 (Al,Si)_8O_{22}(OH)_2$	5.5	Dark mineral, may be greenish, often elongated crystals, in basic igneous rocks and some meta-morphic rocks, two cleavages at about 60°

Andalusite Al_2SiO_5	7.5	Red, often altering to silvery mica. May form cross-shaped crystals. Occurs only in metamorphic rocks. Polymorphs of the same composition are kyanite and sillimante
Anhydrite $CaSO_4$	3–3.5	Forms pale crystals, often breaking into rectangular fragments
Apatite $Ca_5(PO_4)_3(F,Cl,OH)$	5	The most common phosphate mineral. Crystals can take a variety of forms
Biotite (mica) $K(Mg,Fe)_3AlSi_3O_{10}(OH)_2$	2	Virtually black, one excellent cleavage, so that it occurs as fine flakes in igneous and metamorphic rocks
Calcite $CaCO_3$	3	Usually whitish, sometimes good crystals, three directions of cleavage (not at 90°), reacts with weak hydrochloric acid (producing bubbles of carbon dioxide)
Chlorite $(Mg,Al,Fe)_6(Si,Al)_4O_{10}(OH)_8$	2.5	Usually too fine to see, but gives green sheen to fine-grained metamorphic rocks
Epidote $Ca_2(Al,Fe)Al_2(SiO_4)(Si_2O_7)(O,OH)_2$		Usually green elongated crystals, in metamorphic rocks

Name	Hardness	Distinguishing properties
Feldspar $KAlSi_3O_8$ to $NaAlSi_3O_8$ (alkali feldspar) or $NaAlSi_3O_8$ to $CaAl_2Si_2O_8$ (plagioclase feldspar)	6	Usually white or pinkish crystals, two poorly-developed cleavages roughly at 90°
Fluorite (also known as fluorspar) CaF_2	4	Colourless to pale purple or orange, normally occurs as cubes in veins associated with ore minerals, four excellent cleavage directions
Garnet $(Ca,Fe,Mg)_3Al_2(SiO_4)_3$	7	Red, brown or green, no cleavage, most commonly as well-formed crystals in metamorphic rock
Gypsum $CaSO_4.2H_2O$	2	Usually whitish, can occur as good crystals, or as finely crystalline alabaster
Illite $K_yAl_4Si_{8-y}Al_yO_{20}(OH)_4$	n/a	Illite is the most common sort of clay in mudrocks. In the formula given, the value of y is usually in the range 1 to 1.5
Kaolinite $Al_2Si_2O_5(OH)_4$	n/a	A clay mineral with a comparatively simple chemical formula. This is the main component of china clay

Kyanite Al_2SiO_5	4–7	Light blue, bladed crystals that may form rosettes. Occurs only in metamorphic rocks. Polymorphs of the same composition are andalusite and sillimanite
Montmorillonite $(Na,Ca)_{0.3}(Al,Mg,Fe)_4Si_4O_{10}(OH)_2.n\,H_2O$	n/a	A clay mineral that can absorb several times its own weight of water, as implied by n H_2O in its formula
Muscovite (mica) $KAl_2(AlSi_3O_{10})(OH)_2$	2	Usually colourless, one excellent cleavage, so that it occurs as fine flakes in igneous, metamorphic and sedimentary rocks
Olivine $(Fe,Mg)_2SiO_4$	6.5	Pale olive to very dark green, no cleavage
Pyroxene $(Ca,Mg,Fe)_2Si_2O_6$ (augite, diopside) $(Mg,Fe)_2Si_2O_6$ (enstatite)	6	Common dark mineral in basic igneous rocks, two cleavages at about 90°
Quartz SiO_2	7	Colourless or white, but may be tinted by impurities, no cleavage but sometimes shows nice crystal faces
Serpentine $Mg_3Si_2O_5(OH)_4$	3–4	Has various polymorphs, including crysotile, which is fibrous and forms asbestos. A product of hydrous alteration of olivine, pyroxene and amphibole

Name	Hardness	Distinguishing properties
Sillimanite Al_2SiO_5		Long needle-shaped brown or grey crystals. Occurs only in meta-morphic rocks. Polymorps of the same composition are andalusite and kyanite
Talc $Mg_3Si_4O_{10}(OH)_2$	1	When occurring in bulk talc is described as 'soapstone'. A product of hydrous alteration of olivine, pyroxene and amphibole
Zircon $ZrSiO_4$	7.5	An accessory (rare) mineral in igneous rocks, which owing to its hardness is very robust, survives transport well and can be found in sedimentary and even some metamorphic rocks

Table B Table of common rock-forming minerals (all these have a glassy lustre).

Name	Hardness	Distinguishing properties
Cassiterite SnO_2 (tin oxide)	6	Black tetragonal crystals, more often found in water-washed sediments than within hard rock
Chalcopyrite $CuFeS_2$ (copper-iron sulfide)	3.5	Yellow, less brassy than pyrite, tendency to tarnish
Galena PbS (lead sulfide)	2.5	Lead-grey in colour, three good cleavage directions, may occur as cubes
Hematite Fe_2O_3 (iron oxide)	6	Black to dark red in colour, no cleavage, may occur in kidney-like shapes
Pyrite FeS_2 (iron sulfide)	6	Brassy yellow in colour, often occurs as cubes, no cleavage. (This is 'fool's gold')
Sphalerite (zinc blende) ZnS	4	Brown, black or red, tetrahedral crystals

Table C Table of common ore minerals (all these have a metallic lustre).

GEOLOGY

Name	Formula	Notes
Amber	–	Arguably not a mineral. This is fossilized resin
Amethyst	SiO_2	Violet variety of quartz, coloured by impurities, notably iron
Diamond	C (pure carbon)	Formed under high pressure, this is the hardest known mineral
Emerald	$Be_3Al_2Si_6O_{18}$	A form of beryl, coloured green by minute traces of chromium
Garnet	$(Fa,Mg)_3Al_2(SiO_4)_3$	Magnesium-rich varieties are deep red and iron-rich varieties are usually violet
Jet	–	A lustrous variety of coalified fossilized wood (lignite)
Lapis-lazuli	$(Na,Ca)_{4-8}(AlSiO_4)_6(SO_4,S)_{1-2}$	Lapis-lazuli is an ornamental stone, noted for its attractive blue colour, consisting of a fine mixture of lazurite with calcite, pyroxene and other minerals
Opal	$SiO_2.nH_2O$	Amorphous, non-crystalline, hydrated silica
Ruby	Al_2O_3	A form of corundum, coloured red by traces of chromium
Sapphire	Al_2O_3	A form of corundum, coloured blue by traces of titanium and iron
Topaz	$Al_2SiO_4(F,OH)_2$	May be yellow, blue, or even red

Table D Table of notable precious and semiprecious stones.

APPENDIX 2:
ROCK NAMES

No two rock bodies, or, for that matter, no two specimens of rock collected from the same body, have *exactly* the same composition. Even closely similar specimens will differ imperceptibly in their chemical ingredients, or in the size or shape of the crystals or grains that make up the rock. However, a basalt and a granite are so clearly different in appearance and origin that it is both useful and sensible to have distinct names for them. This appendix lists the more common rock names and the criteria used to distinguish them.

Igneous rocks

These are rocks that formed from a molten state. Most igneous rocks are composed of an interlocking collection of crystals that grew as the melt (magma) cooled. The crystals usually have a random orientation, and, generally speaking, the slower the magma cooled, the larger the crystals. Consequently, most igneous rocks can be identified as such simply on their basis of their textures. Coarse-grained igneous rocks (average crystal size greater than 2 mm across) can be identified with the unaided eye, medium-grained igneous rocks (average crystal size between 0.25 mm and 2 mm) usually need a hand lens or magnifying glass, and fine-grained igneous rocks (average crystal size less than 0.25 mm) require a microscope to be certain of their nature and composition.

There are several notable exceptions to these general remarks: (i) flow or settling during crystallisation can cause crystals to become oriented in layers, (ii) some volcanic igneous rocks may have cooled so rapidly that no crystals had time to grow, with the result that the rock has the appearance, and sub-microscopic structure, of glass, (iii) volcanic rocks that were erupted explosively (pyroclastic rocks) are made of either glassy or crystalline fragments, (iv) the very coarse grain size of rocks described as pegmatites (see Chapter 6) bears no relation to the rate of cooling.

Whatever their texture, igneous rocks are formally classified on the basis of their chemical composition, notably by the proportion of silica (SiO_2) in their total chemical make up. Chemical analyses cannot usually be made in the field, but the overall chemistry of a rock is reflected in the kinds of minerals that it contains, so fortunately it is possible to identify rock types in this way. The more silica a rock has, the more acidic it is said to be, and igneous rocks are divided into acidic, intermediate, basic, and ultrabasic types.

Acid igneous rocks

These contain greater than 66% silica (SiO_2), which is not to say that they contain greater than 66% quartz (the mineral with SiO_2 as its formula), because all silicate minerals contain some SiO_2 in their formulae. The most abundant minerals in an acidic igneous rock are alkali feldspar, quartz, and plagioclase feldspar, with lesser amounts of muscovite, biotite and sometimes amphibole. If the rock is coarse grained it is called **granite**, if it is medium grained it is called **microgranite**, and if it is fine grained it is called **rhyolite**. The term 'granitic' is used to describe acidic igneous rocks in general; for example a rhyolite is a fine-grained acidic igneous rock. A pyroclastic rock of this composition would be referred to as a 'rhyolitic tuff'.

Intermediate igneous rocks

These contain between 52% and 66% silica. The most abundant mineral is plagioclase feldspar, with lesser amounts of biotite and amphibole, and sometimes alkali feldspar, quartz, or pyroxene. A coarse-grained intermediate igneous rock is called **diorite**, its medium-grained relative is called **microdiorite**, and the fined-grained equivalent is called **andesite**. The term 'andesitic' is used to describe these rocks in general, so that a diorite can be said to be a coarse-grained andesitic rock.

Basic igneous rocks

These contain between 45% and 52% silica (but the mineral quartz is entirely absent). The most abundant minerals are plagioclase feldspar and pyroxene, and sometimes olivine. A coarse-grained basic igneous rock is called a **gabbro**, its medium-grained equivalent is called a **dolerite** (British usage) or **diabase** (American usage), and the fine-grained variety is called **basalt**. The term 'basaltic' is used to describe these rocks in general.

Ultrabasic igneous rocks

These contain less than 45% silica (usually not reaching less than about 38%). The most abundant mineral is olivine, with pyroxene and a lesser amount of plagioclase feldspar. Only the coarse-grained variety is at all common, and is named **peridotite**.

Other distinctions

The names highlighted above are just the most common names, and there is a bewildering variety of names still in use. Some of these are inherited from the days when each rock type was named after the place it was found (for example, a certain variety of basalt is still referred to as 'hawaiite', after Hawaii). Other names recognise that silica content is not the only significant variable in rock composition: another is the relative abundance of the 'alkali' metals sodium and potassium (which tend to go into alkali feldspar) compared to calcium (which goes into plagioclase feldspar). Rock names such as syenite, monzonite and dacite reflect a decreasing abundance of alkalis, in alkali-rich rocks that are broadly of granitic to andesitic composition.

Metamorphic rocks

These are rocks that form when heat or pressure, or both, cause a pre-existing rock to recrystallise. Very often, the effect of pressure is to cause crystals to grow in, or to re-orient themselves into, layers. When this happens the metamorphic rock has a layered texture, which described as 'foliation'. In fine-grained metamorphic rocks the crystals cannot be seen by the unaided eye, but the foliation is manifested by the tendency of the rock to fracture along flat or wavey 'cleavage' surfaces parallel to the foliation.

Metamorphic rocks are most easily classified simply on textural grounds, using the same fine, medium and coarse grain size division as in igneous rocks, as shown in the following table.

Textural type	Nature of foliation	Grain size
Slate	Very closely spaced, almost flat	Fine
Phyllite	Wavey or crenulated, sheen caused by minute flakes of mica or chlorite on foliation surfaces	Fine
Schist	Undulating planes, characteristically with abundant mica	Medium to coarse
Gneiss	More widely spaced than in schist, usually alternating layers of dark and light minerals	Coarse
Migmatite	Poor foliation, shows signs of having begun to melt	Coarse

Table X Classification of metamorphic rocks by texture.

It is often convenient to qualify the simple terms listed above in recognition of prominent minerals within the rock. For example a schist with abundant mica and distinct crystals of garnet may be described as a garnet mica schist.

Not all metamorphic rocks have a foliated texture. For example a rock metamorphosed by the heat of an igneous intrusion (contact metamorphism) usually develops a spotty texture and is described as a **hornfels**. Furthermore, rocks that, prior to metamorphism, consisted of just one abundant mineral are unlikely to develop a new mineral assemblage during metamorphism. The original texture may be destroyed by adjustment of crystal boundaries, but there will not be a strong oriented fabric. A quartz-rich sandstone that has been metamorphosed consists of interlocking quartz crystals and is described as **quartzite**, and a limestone that has been metamorphosed consists of interlocking calcite crystals and is described as **marble**.

Chapter 7 introduced the concept of metamorphic facies, corresponding to the temperature and pressure under which metamorphism took place. The facies cannot usually be determined by simple inspection;

however, when the facies is known then it is normal to write of (say) an amphibolite facies gneiss or a zeolite facies slate.

Sedimentary rocks

These form by deposition of detrital grains or by precipitation from solution in water. Detrital (or 'fragmental') sediments are most simply distinguished from igneous and metamorphic rocks on textural grounds, because the grains of which they are composed are non-interlocking fragments rather than interlocking crystals. Sediments formed by precipitation do have interlocking cystalline textures, but are made of quite different minerals to igneous rocks, notably calcite (in limestone) or evaporite minerals such as gypsum.

Fragmental sediments are classified according to grain size in the scheme shown in the following table, which shows names for the rock type and the different sizes of clasts (n.b. breccia is pronounced 'brechia').

Grain size	Term for clasts	Rock type
>256 mm	boulders	**Conglomerate** (rounded clasts) or **breccia** (angular clasts)
64–256 mm	cobbles	**Conglomerate** (rounded clasts) or **breccia** (angular clasts)
2–64 mm	gravel (pebbles 4–64 mm)	**Conglomerate** (rounded clasts) or **breccia** (angular clasts)
0.0625–2 mm	sand	**sandstone**
0.002–0.0625 mm	silt	**siltstone**, or **mudrock** (referred to as **shale** if it splits easily into layers)
<0.002 mm	clay	**claystone**, or **mudrock** (referred to as **shale** if it splits easily into layers)

Table Y Classification of sedimentary rocks by grain size.

Rocks of these compositions may sometimes also be described according to their supposed mode of origin (which is a risky business, because the geologist may get this wrong!). For example, a breccia or conglomerate with a muddy matrix that is thought to have been deposited by a glacier would be called a tillite.

Sandstones are further classified according to the mineralogy of their grains. A sandstone that is over 90% quartz grains is described as a quartz sandstone, whereas one containing a significant amount of feldspar, clay minerals or calcite is called an arkose, muddy sandstone or calcarous sandstone, respectively. One that has a significant proportion of fragments composed of fine-grained rocks is described as a greywacke.

A rock that consists of over 90% calcite is referred to as a **limestone**. A limestone with an admixture of quartz grains or clay minerals would be called a sandy limestone or a muddy limestone. Pure limestones are further distinguished according to whether the calcite occurs as fragments of shells (skeletal limestones), or as ooids (oolitic limestone), and whether the grains are cemented together by crystalline calcite (sparite) or held together by fine-grained calcite mud (micrite). If the calcite has been replaced by the mineral dolomite, the rock should be called a dolostone, but is more commonly referred to as **dolomite**.

Another common sedimentary rock name is **chert**, which describes a rock formed of microscopic silica crystals (which may be a chemical precipitate, as in the case of flint, or shells of microscopic organisms such as radiolarians).

APPENDIX 3:
GEOLOGIC TIME

Geological time is divided according to the scheme shown in the accompanying diagram. Note that this is *not* drawn to a uniform scale. The numbers refer to the age in millions of years of the boundary between each division, as determined radiometrically (i.e. by measuring the products of radioactive decay). These dates are still subject to revision.

The largest unit of geologic time is an Eon. The scheme shown here names four Eons. The oldest known rocks date back to the beginning of the Archean, and the Hadean is that time before for which we have only indirect evidence. Some authorities do not regard the Hadean as a separate Eon, and include this time within the Archean.

The Phanerozoic Eon is divided into three Eras – Paleozoic, Mesozoic, and Cenozoic (which means old-life, middle-life, and modern-life respectively). In many kinds of sedimentary rocks of Phanerozoic age fossils are sufficiently abundant and show sufficiently clear evolutionary change to enable geological time to be split into a succession of progressively smaller divisions. Each Era is divided into a number of Periods. Each Period is divided into a number of Epochs (shown here only for the Paleogene and Neogene Periods), and further subdivided into Ages (not shown here).

The Holocene Epoch covers the past 10 000 years, and is sometimes called the Recent. The time span covered by the Cenozoic was formerly split into two Eras – the Tertiary (from the start of the Paleocene to the end of the Pliocene) and the Quaternary (Pleistocene and Holocene), and both these terms are still in widespread use. The boundary between the Cretaceous and the Tertiary, which is marked by the most recent major episode of mass extinction, is often referred to as the K/T boundary (K being the usual abbreviation for Cretaceous).

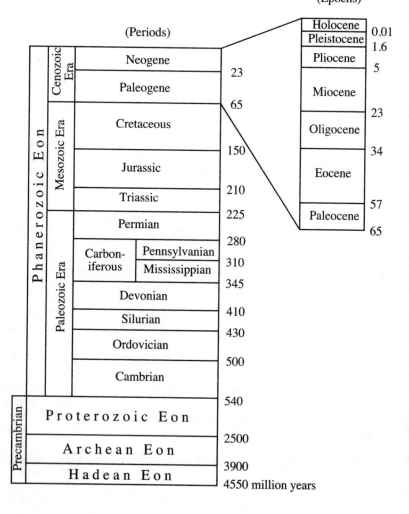

GLOSSARY

This Glossary contains those terms that I think the newcomer to geology will have most trouble with. I have indicated the chapter(s) in which each term makes an important appearance. Names of several minerals and rock types not listed here may be found in Appendices 1 and 2. There is a stratigraphic timescale in Appendix 3.

Andesite a fine-grained igneous rock type richer in silica than basalt, commonly erupted at volcanoes above subduction zones (Ch 5)

Anticline an arch-like fold, in which beds on either side dip away from each other (Ch 10)

Ash fine, fragmentary rock, produced by a pyroclastic (explosive) eruption (Ch 5)

Asthenosphere the weak zone within the mantle (but below its top), which is overlain by the lithosphere (Ch 2)

Banded iron formation, BIF a sedimentary deposit consisting of alternate layers of iron oxide and silica, of the kind constituting most of the global iron reserves. Most are between 3800 and 1800 million years old (Ch 11)

Basalt an igneous rock type having the same composition as oceanic crust. Strictly speaking, a basalt must be fine grained; coarser grained rocks of the same composition are described as basaltic (Ch 2, 4)

Bedform a ripple, dune or other similar feature produced by flowing water or wind (Ch 9)

Bed-load particles dragged, rolled or bounced along during transport by wind or water (Ch 8)

Cement the precipitated mineral holding grains together in a sedimentary rock, most commonly calcite or silica (Ch 9)

Clast a particle or lump (of any size) that is transported and deposited in a sediment (Ch 9)

Clastic rock a sedimentary rock composed of clasts (Ch 9)

Cleavage regular planes of weakness in a crystal, which are a consequence of its atomic structure (Ch 6, 8). The same term describes closely spaced planes of weakness in a deformed or metamorphosed rock, which are caused by alignments of platey minerals (Ch 7, 10)

Constructive plate boundary a site where two tectonic plates are moving apart, and each is being added to by the formation of new oceanic lithosphere (Ch 4)

Continental drift the relative movement of continents, which is a consequence of plate tectonics (Ch 4)

Core in the Earth, high-density iron-rich interior (3500 km in radius). It consists of a liquid outer core and a solid inner core (Ch 2)

Craton a stable area of continental crust, also known as a shield, that has remained undeformed since the Archean (Ch 13)

Crust in the Earth, the compositionally distinct, slightly lower density, rocky layer overlying the mantle (Ch 2)

Destructive plate boundary a site where one tectonic plate is destroyed by subduction below another (Ch 4)

Diagenesis the chemical and physical processes that turn an unconsolidated sediment into a hard sedimentary rock (Ch 9)

Dip the angle between the horizontal plane and a geological surface, such as a bedding plane (Ch 15)

Dyke a sheet of igneous rock, intruded vertically (Ch 6)

Epicentre the point on the Earth's surface directly above an earthquake's focus (Ch 3)

Evaporite a mineral or deposit of minerals formed by evaporation of seawater (Ch 9)

Facies for metamorphic rocks, the temperature and pressure conditions under which they were metamorphosed (Ch 7)

Focus the break-point where earthquake motion begins, and where most of the energy is released (Ch 3)

Foliation an alignment of minerals in planes, caused by regional metamorphism (Ch 7)

Fossil the remains of a past organism (Ch 12)

Fossil fuels coal, oil and natural gas, which are being used up at rates much faster than the rates at which they are being created by geological processes (Ch 11)

Fractional crystallisation the process whereby the first crystals to form in a magma as it cools have a composition different to that of the magma as a whole (Ch 6)

Granite a coarse-grained intrusive igneous rock, richer in silica than an andesite. Rocks of this composition are described as granitic (Ch 5, 6)

Greenhouse effect the warming of a planet's atmosphere by means of absorption of infra-red radiation by atmospheric gases (Ch 13)

Hornfels a splintery, spotty rock produced by contact metamorphism (Ch 7)

Hot spot a place where a plume rising from deep within the mantle hits the base of the lithosphere, giving rise to large volumes of basaltic magma (Ch 5)

Ice Age a period (of the order of ten million years long) when the climate is cold and polar ice sheets are particularly extensive (Ch 8, 13)

Igneous referring to a rock or mineral formed by solidification from a molten state (Ch 1)

Intrusive referring to an igneous rock emplaced below the surface (Ch 6)

Island arc a usually arcuate line of islands built by volcanism above a subduction zone (Ch 4)

Lava molten rock at the surface. The term is also used to describe rock formed in this way after it has solidified (Ch 5)

Lithosphere the rigid outer shell of the Earth, consisting of the crust and the uppermost mantle (Ch 2)

Magma molten rock; usually used to refer to molten rock at depth, whereas molten rock at the surface is called lava (Ch 4)

Mantle in the Earth, all the silicate (rocky) material surrounding the core, with the exception of the thin silicate crust (Ch 2)

Metamorphic referring to a rock or mineral formed by recrystallisation, without melting but in response to heat and/or pressure, from a pre-existing rock or mineral. This process is called metamorphism (Ch 1, 7)

Mineral a naturally occurring crystalline substance with a well defined chemical composition. Most rock types consist of several different minerals (Ch 6). See Appendix 1 for a fuller discussion

Moho the boundary between the crust and the mantle (Ch 2)

Non-renewable resources fossil fuels, ores and bulk materials, which are being used up at rates much faster than the rates at which they are being created by geological processes (Ch 11)

Normal fault steep fault with vertical displacement, where the rock on the down-dip side of the fault has moved downwards (Ch 10)

Ophiolite a slice of oceanic crust and upper mantle that has been thrust over the edge of a continent (Ch 4)

Ore a metal-bearing mineral from which the metal can be extracted at a profit (Ch 11)

Orogeny a mountain-building episode, as a result of a collision between two continents, or between a continent and an island arc (Ch 13)

Partial melting what happens when any rock composed of a variety of minerals begins to melt. The first melt to form has a different chemical composition (richer in silica) than the solid (Ch 4)

Permeability the ease with which fluid or gas can flow through a porous rock, which is dependent on how well connected the pore-spaces are (Ch 11)

Peridotite a rock type poor in silica, having the same composition as the Earth's mantle (Ch 2, 4)

Phenocryst a larger crystal surrounded by smaller crystals in an igneous rock (Ch 6)

Pillow lava the characteristic morphology shown by basaltic lava when erupted under water (Ch 5)

Placer deposit an ore concentrated in a sediment, because the ore forms the densest grains (Ch 11)

Plate tectonics a description of how the rigid plates into which the Earth's lithosphere is divided move about (Ch 4)

Porosity the proportion of a rock body that is occupied by liquid- or gas-filled spaces; contrast permeability (Ch 11)

Prograding description of a shoreline that is building out seawards (Ch 9)

Pyroclastic term used to describe an eruption (or the resulting rock) in which fragments are produced by explosive volcanic activity (Ch 5)

Radioactive (radiogenic) heating heating of a planet's interior through the energy released by decay of unstable, radioactive atoms (Ch 2)

Radiometric dating a technique in which the relative proportions of products of radioactive decay and their parent isotopes are used to determine the age of a rock or mineral (Ch 2)

Reserves those resources that can be extracted profitably and legally under existing conditions; contrast resources (Ch 11)

Reservoir rock a permeable body of rock into which hydrocarbons or water have migrated, and from which they can be extracted through a well; contrast source rock (Ch 11)

Resources the estimated total amount of a commodity (fossil fuel, ore or raw material) available in the world; contrast reserves (Ch 11)

Reverse fault steep fault with vertical displacement, where the rock on the down-dip side of the fault has moved upwards (Ch 10)

Sedimentary referring to a rock formed from a deposit of detrital clasts, or by precipitation from solution in water. Sedimentary minerals are those that form in, or are found in, sedimentary rocks (Ch 1)

Seismic waves vibrations, from earthquakes or artificial explosions, that travel through the Earth. P-waves are compressional waves, and S-waves are shearing waves (Ch 2)

Seismometer device for recording the vibrations caused by earthquakes or movement of magma (Ch 2, 3, 5)

Shale a mudrock or siltstone that breaks naturally into finely-spaced layers parallel to its bedding (Ch 11, Appendix 2)

Shield another world for craton (Ch 13)

Shield volcano a gently-sloping basaltic volcano, so named because of its cross-sectional profile (Ch 5)

Silica any compound with the formula SiO_2. The term is also used to refer to the percentage of a rock (by weight) that can be expressed as SiO_2 irrespective of whether this occurs as pure silica or combined with other elements in silicate minerals (Ch 2)

Silicate either a rock type or a mineral rich in silicon and oxygen. The Earth's mantle and most of its crust are formed of silicates, as are virtually all igneous rocks (Ch 2, 6)

Sill a horizontal sheet-like igneous intrusion (Ch 6)

Source rock a body of rock within which hydrocarbons form; contrast reservoir rock (Ch 11)

Strike the direction, conventionally measured clockwise from north, of a horizontal line drawn on a geological surface, such as a bedding plane (Ch 15)

Subduction term describing one tectonic plate descending at an angle below another, which happens at a destructive plate boundary (Ch 4)

Suspension-load particles suspended for long periods without touching the bottom during transport by wind or water (Ch 8)

Syncline a U-shaped fold, in which beds on either side dip towards each other (Ch 10)

Thrust a low-angle fault which has pushed older rocks over younger rocks (Ch 10)

Transcurrent fault a steep fault across which relative displacement is sideways. Alternatively known as a strike-slip fault (Ch 10)

Transform fault a steep fault offsetting two lengths of a constructive plate boundary, which has sideways slip across it (Ch 4)

Tsunami a water wave triggered by an earthquake, landslide or volcanic eruption. The wave's height is slight while it travels through deep water, but it becomes much higher and extremely dangerous as it approaches the shore (Ch 3, 5)

Ultrabasic term describing rock that has even less silica than a basalt, such as a peridotite (Ch 6)

Unconformity a boundary between underlying older rocks and an overlying set of younger strata, usually manifested by a difference in dip and representing considerable time gap (Ch 10)

Vein a fracture that has become filled by minerals precipitated from solution, commonly quartz or calcite, and sometimes including ore minerals (Ch 11). Coarse veins described as pegmatite may crystallise from the last fraction of melt remaining as an igneous intrusion cools down (Ch 6)

INDEX

Note: many minerals and gemstones not listed here can be found on pages 211-218, and some more rock names occur on pages 220-224.